ALONE BUT NOT LONELY

EXPLORING FOR EXTRATERRESTRIAL LIFE

LOUIS FRIEDMAN

FOREWORD BY MAE JEMISON

———————————————

———————————————

ALONE BUT NOT LONELY

Exploring for Extraterrestrial Life

THE UNIVERSITY OF
ARIZONA PRESS

TUCSON

The University of Arizona Press
www.uapress.arizona.edu

We respectfully acknowledge the University of Arizona is on the land and territories of Indigenous peoples. Today, Arizona is home to twenty-two federally recognized tribes, with Tucson being home to the O'odham and the Yaqui. Committed to diversity and inclusion, the University strives to build sustainable relationships with sovereign Native Nations and Indigenous communities through education offerings, partnerships, and community service.

ISBN-13: 978-0-8165-4950-4 (paperback)
ISBN-13: 978-0-8165-4951-1 (ebook)

Cover design by Leigh McDonald
Cover photo: This "Einstein Ring" is a gravitational lens image of a distant galaxy. Using gravitational lens magnification to image exoplanets is how we can explore for extraterrestrial life. Image by ESA/ Hubble & NASA, S. Jha, Acknowlegment: L. Shatz
Designed and typeset by Leigh McDonald in Arno Pro 11.5/14 (text) and Alternate Gothic Compressed (display)

Library of Congress Cataloging-in-Publication Data
Names: Friedman, Louis, author.
Title: Alone but not lonely : exploring for extraterrestrial life / Louis Friedman.
Description: Tucson : University of Arizona Press, 2023. | Includes bibliographical references and index.
Identifiers: LCCN 2022051877 (print) | LCCN 2022051878 (ebook) | ISBN 9780816549504 (paperback) | ISBN 9780816549511 (ebook)
Subjects: LCSH: Life on other planets.
Classification: LCC QB54 .F925 2023 (print) | LCC QB54 (ebook) | DDC 576.8/39—dc23/eng20230508
LC record available at https://lccn.loc.gov/2022051877
LC ebook record available at https://lccn.loc.gov/2022051878

Printed in the United States of America
♾ This paper meets the requirements of ANSI/NISO Z39.48-1992 (Permanence of Paper).

CONTENTS

ILLUSTRATIONS

FIGURES

TABLES

FOREWORD

THE MYSTERY OF space has permeated human inquiry since we first looked up. And being the tenacious, curious creatures we are, we explored. Through observation—recognizing movement and patterns—we calculated and sorted possibilities of what might make up the heavens that was consistent with the multitude of observations. This happened across the world, in every culture and society. As time passed and civilizations became more sophisticated, we applied our technologies and capabilities to explore and refine our "knowledge"—physically and mentally. And it was always about prognostication—prediction of what will come or may be.

Prognostication is difficult. It is fraught with pitfalls and stepping stones dependent on the prognosticator's origin story—their experiences, culture, learnings, knowledge base, training, and access, all of which are influenced by conscious and unconscious intellectual and emotional leanings.

In *Alone but Not Lonely*, Dr. Friedman takes a courageous dive into the mystery of the sky above us and endeavors to divine our place and future in this universe and whether we will ever meet our neighbors, face to face, in person.

His is a major, admirable, far-reaching undertaking to corral current knowledge of space—from probes and increasingly powerful telescopes to astrophysicsal theories and rocket science—and combine it with growing knowledge of biological systems on Earth, and then predict the likelihood of humans ever traveling in person to other worlds outside our solar system or finding other beings not from Earth with whom we will be able to communicate.

Alone but Not Lonely is a worthwhile, authoritatively written, engaging read. I do not align with all his conclusions; still I believe that this work focuses our attention and provokes us to dig deeper and respond.

Dr. Friedman's exploration of the operations and results to date of searches for extraterrestrial intelligence (SETI) through radio astronomy is clear and accessible to everyone. Friedman jumps courageously into the life sciences and gives his understanding on intelligence and evolution of life as documented so far. The practical take on interstellar travel is grounded in what we are capable of today and technologies currently on the horizon. He explains the possibility of using the sun's gravitational bending of light to create a powerful lens to view planets in other solar systems up close—allowing us to potentially explore not only the geological features of planets but also the life that may have evolved on planets orbiting other stars: astrobiology without making the physical journey.

Dr. Friedman and I share a fascination for and commitment to exploring space beyond our solar system. We have held conversations, joined in thought exercises and combined efforts to build paths to expand human access to and understanding beyond our Sun's influence deep into interstellar space and our galaxy.

I lead the 100 Year Starship° global initiative that seeks to ensure that the capabilities for the human interstellar journey exists within the next hundred years. Why? Because including the physical presence of humans in an interstellar journey forces us to think *way* outside the box, across the full spectrum of technologies, human systems, knowledge, and experience. While we obviously have intersecting knowledge, Dr. Friedman's and my backgrounds are different in disciplines, experience, and knowledge base. As an engineer, medical doctor working in extreme conditions, environmental studies professor, social scientist, and astronaut, I am certainly familiar with and share Dr. Friedman's embrace of the practical. However, I depart from him as I also recognize though, while seemingly familiar and commonplace, how little we really understand about life, its habits, and its origins. Many frequently mistake the familiar and commonplace predictability of Newtonian physics as the brackets for understanding, identifying, and assessing the attributes, capabilities, and possibilities of both life and intelligence.

The beauty of this book is that it provides a perspective that challenges us to think more deeply. I found myself having a conversation with the text, providing evidence for and contrary to various assumptions and arguments that would lead to different conclusions. *Alone but Not Lonely* offers new possibilities for the future while seeming to preclude others. But I'll stop now, because the prize here is in the reading.

Keep in mind while reading, though, that famed science fiction author and scientist Arthur C. Clark said, "If an elderly but distinguished scientist says something is possible, he is most certainly right; but if he says that it is impossible, he is very probably wrong."

In a universe as vast in possibilities as it is in physical dimensions, we must be vigilant in acknowledging that we have sampled only a miniscule portion. While our knowledge database has expanded and continues to evolve rapidly beyond what we knew even twenty years ago, we are still a young species—even in Earth terms. We have much more to experience and learn. I consciously attempt not to allow my limited observations to limit my search for the possible outcomes. Space exploration is not about practicality—for me, it is the ultimate in possibility. This book piques my commitment to explore ever further.

Mae Jemison, M.D.

ACKNOWLEDGMENTS

Sir Isaac Newton said, "If I have seen further than others, it is because I stand on the shoulders of giants," modestly stating that his genius work owed a debt to many mathematicians and scientists before him. My work of course is not genius, but it too stands on the shoulders of a few giants and many of those of lesser height who contributed great things to whatever I have learned and understand. These range from a brilliant high school math teacher (Dr. Julius Hlavaty) to my PhD mentors, Wallace Van Der Velde, Dick Battin, and Advisor Irwin Shapiro, to Carl Sagan and Bruce Murray, with whom (in a fantastic and fortuitous association) we created The Planetary Society (with many other colleagues and friends too numerous to name). Our motto (or, as they say now, our "mission statement") was "to explore new worlds and seek other life." I was extremely fortunate to be able to work with brilliant people at the Society for thirty years and to participate in such exploration and seeking.[1] Two of them were Arthur Clarke and Isaac Asimov. Clarke's famous "Either we are alone in the universe, or we are not ... either prospect is terrifying," and Asimov's seminal work with the "three laws of robotics" shape much of what is in this book. Those three laws are, briefly: (1) do no harm to humans; (2) obey humans unless it contradicts the first law; and (3) protect yourself (the robot) unless it contradicts the first two laws. These laws seem straightforward until we bring artificial intelligence and deep learning into our robots.

A robot that can think would perhaps be allowed to break Asimov's laws. In this book I will speculate on the evolution of the human species, with one possible outcome being a robotic/AI symbiosis—a potential new species that might compare to humans as humans compare to Neanderthals. This subject comes up later.

Another completely different giant I met at The Planetary Society was John Gardner, the founder of Common Cause. He noted that political reform was a bit like science—you are never done, you just have to keep working at it. That advice was in the context of creating a public-interest advocacy organization, but it has a special insight when applied to science. Science may seek answers, but it's more about asking questions. That is very important in this book. I will assert much, aiming to provoke questions about the nature and role of life in the universe. Is it unique or ubiquitous, is it the inevitable product of physical and chemical evolution, or is it an accident of random perturbations and chaotic events, destined for progress or for extinction? We don't know, but we'll just keep working at it.

In addition to Carl Sagan and Bruce Murray, who opened my eyes to the questions about life beyond Earth, I want to acknowledge my many colleagues and friends at The Planetary Society who helped me learn about other worlds and extended my Planetary Adventures (the title of my previous book) to many planets and promoted the idea of humans becoming a multiplanet species. In connection with this book, I particularly thank Slava Turyshev, who is developing the idea of the solar gravity lens as Nature's telescope for us to use, and Darren Garber, who generously provided the astrodynamics technical analyses that will permit us to reach the solar gravity lens and possibly do a lot more on the way out. Dr. Turyshev also graciously helped review this book for errors—any remaining ones are mine. Thank you also to Allyson Carter and the folks at the University of Arizona Press for their interest and advice leading to the book's publication. And a special thank-you to astronaut and doctor Mae Jemison for her foreword to this book and for the inspiration and vision she provides for a positive future.

Finally, I particularly acknowledge my wife, Connie, who passed away while I was writing this book after providing me with fifty-six years of support and love.

ALONE BUT NOT LONELY

EXPLORING FOR EXTRATERRESTRIAL LIFE

INTRODUCTION

Why write another book about extraterrestrial life? There are many already—probably many thousands: fiction, nonfiction, anthologies, religious, comic, pseudoscience, texts, articles in encyclopedias, etc. A lot for a subject that still has no subject matter. (That is, there is yet no discovery and no knowledge of extraterrestrial life.) The answer is because it is interesting. And why is it interesting? One answer is because it is an unknown—a mystery wanting to be solved. Another is because it is important: understanding life in the universe is fundamental knowledge about ourselves—who and what we are, how and when we came to be, and most importantly why. If you are not interested in whether life in the universe is unique (just us) or ubiquitous, then (as my colleague, mentor, and friend Carl Sagan used to say) you must be made of wood.[1] The interest in extraterrestrial life throughout human history is the subject of chapter 1.

Ok—that is the case for one more book. But why one by me? After fifty-plus years in space, that is, in the space business on both the technical side and the public interest side, I have come to three big conclusions about extraterrestrial life: (1) it is the dominant motivation and rationale for space exploration; (2) the universe is likely teeming with life and waiting for us to explore; (3) but extraterrestrial intelligence and interstellar travel are more subjects of science fiction than they are of science and are likely to remain so. This last one is disappointing—I worked for many years to reach opposite conclusions and actively supported Search for Extraterrestrial Intelligence (SETI) projects and participated in interstellar flight studies. I am on the Starshot Advisory Committee, whose goal is to privately fund the development of an interstellar flight, and I participated in the start-up of The 100 Year Starship, a group now devoted to advancing interdisciplinary research related to interstellar exploration and life on other worlds.[2] Interstellar flight and SETI are related, both dealing with projecting ourselves into the universe for contact with others, and ultimately they both fail for the same reasons—the universe is just too big and our lifetime is too short. The first part of the book's title, "Alone," stems from this third conclusion, while the first two conclusions qualify it: "but Not Lonely."

This book is written for a public audience, and most people in that audience conflate the questions of extraterrestrial life with those of extraterrestrial intelligence. Many casual readers are more familiar with SETI than they are with astrobiology (the study of questions related to extraterrestrial life). Whenever I speak publicly about extraterrestrial life (or at parties and in bars) the most common question is "Do you think there is life out there?" followed immediately by "When will we have contact?" The second question implicitly assumes "intelligent life" that can contact or be contacted by us. But if most life are plants or microbes (as it is on Earth[3]), then "contact" isn't the right word; "discovery" is. Chapters 2 and 3 deal with SETI and the questions of extraterrestrial intelligence. I define what I mean by "intelligence" (ignoring the question posed by many, "Is there intelligent life on Earth?") and review the various searches of the past and present and why I conclude that intelligent life in the universe is most likely very rare, maybe even unique to Earth, and maybe even here it is a rather short-lived episode in the long evolution of our planet.

Stop! Bear with me for another minute. I am not a doomsayer or a negative Nellie, nor am I suggesting any depressing result. First, if we are the only intelligent species in the universe, that makes us pretty special and we should certainly try to get things right. Second, as we get to in chapter 4, even if we are the only "intelligence," the universe is also likely teeming with simpler life on billions of planets to explore. Exploration (in my view) is the sum of discovery and adventure, and I propose there will be a lot of adventure in exploring the universe and studying habitability, life, and how the physical universe connects to the biological one.

However, again I lead with what some will say is naysaying: the negative conclusion in chapter 5 is that we cannot physically explore, visit, or even get to other worlds beyond our solar system. There are billions of potentially habitable or even inhabited worlds, but we can't get to them. Again, the universe is too big, and our lifetime is too short. Fortunately, Nature and terrestrial intelligence come up with something far more realistic and more practical. Instead of wishing we could reach one interstellar destination by summoning up huge, impractical rockets, we can explore many interesting ones using Nature's telescope—the solar gravitational lens created by the bending of light by gravity—and our evolving technologies of artificial intelligence and virtual reality (with real exploration). These subjects are dealt with in chapters 7 and 8, respectively. It is not naysaying to say no interstellar flight—it is offering a new, faster, better, more comprehensive, and ultimately more exciting way to explore life on other worlds—virtually with multiple solar gravity lens missions. And, just as exciting, we do this with an army of small telescopes placed way out in the interstellar medium on the focal lines of exoplanets, but still close enough to reach with fast spacecraft. This is the ultimate purpose of this book—to suggest a new way to explore many worlds, to observe what we expect to be many different kinds of life on many different worlds, and to use such observations to investigate the questions of what life is and, with that understanding, to learn more about who we are. It's enough to keep us busy and excited for thousands of years. So, if some science fiction or space enthusiasts think that is not enough (because I downplay the possibilities of contact and interstellar travel), well, come see me after those thousands of years and we will discuss it.

I do not oppose SETI scientific research and observation programs; they are low-cost and result in interesting technological and astronomical discoveries, even if the ultimate desired outcome is unlikely. Similarly, I do not oppose engineering studies of interstellar flight—in fact I am participating now with the Breakthrough Initiatives project Starshot, whose goal is to do just that. Studying difficult science and technology subjects is good, and the research into the technical advances necessary for interstellar flight can be valuable since it often leads to unexpected results in engineering and science even if the goal of interstellar flight remains beyond our reach. In particular I admire the work of the SETI Institute and of Breakthrough Initiatives—brilliant work with smart people chasing elusive goals.

And I may be wrong! It happens. Things once thought to be impossible do come to pass (for example, airplane flight, reaching the Moon, eliminating smallpox, the four-minute mile, tiny computers). That of course doesn't mean there aren't plenty of things that are impossible (such as my going faster than the speed of light or walking on water). But whether I am right or wrong I hope the exposition in this book and the inquiries suggested provoke you and everyone into embarking on more exploration, with more discovery and more adventure.

THE SEARCH FOR EXTRATERRESTRIAL LIFE
—FROM RELIGION TO SCIENCE

Is human life unique in the universe or is it ubiquitous?
A profound and popular question.

THE QUESTIONS ABOUT extraterrestrial life have been with us for all human history: recorded and prehistoric. Some published analyses of prehistoric cave art suggest both the recording of star positions and evidence of timekeeping. There is no question that prehistoric humans and perhaps even Neanderthals observed and wondered about the stars. The ancient Greeks and Romans debated the existence of extraterrestrial life. Democritus, Epicurus, and Lucretius all supported the idea of life on other worlds. Lucretius noted, ". . . you are bound to admit that in other parts of the universe there are other worlds inhabited by many different peoples and species of wild beasts."[1] Aristotle and Plato did not agree with this idea: Aristotle objected on religious grounds, believing that multiple uninhabited worlds were inconsistent with a "prime mover" who created everything. Plato's rejection was based on his finding that human uniqueness was part of "absolutism"—things being just so. This absolutism seems similar to more orthodox or fundamentalist religious views still widespread today, although hardly common.

Many years ago (in the mid-1980s) I was invited to a meeting with the Dalai Lama. I was invited not because of any spiritual connection or philosophical insights, but because of my very secular position as executive director of The Planetary Society. The Dalai Lama was visiting the Center for the Study of Democratic Institutions at the University of California, Santa Barbara, which organized a meeting with leaders of nonprofit groups with goals of spreading democracy through public interest and engagement. The Planetary Society was one of the groups.[2] The Center's president, Harry Ashmore, helped found the Society and was one of our early advisors. The Dalai Lama was very interested in science and astronomy and was especially interested in questions about life in the universe. He asked about the Search for Extraterrestrial Intelligence. I spoke about SETI and he spoke about equal consideration of all life, a view that can be summarized in his statement, "As soon as you look at others as something different from you, and you consider yourself as something different, then this sort of uneasiness comes."[3] This is relevant, not because I share his view or necessarily accept the existence of extraterrestrial intelligence (ETI[4]), but because it underlies for me why I believe SETI to be a worthy effort even while not accepting that ETI even exists. It leads to valuable research and inquiry about the fundamental questions of what life is and what our relationship to the universe is.

Harry Ashmore's presence on our advisory board and his role in The Planetary Society's formation further illustrates the breadth of space exploration and the search for extraterrestrial life. He was a Pulitzer Prize winning journalist from Arkansas whose fame was due to a series of articles he wrote about racial segregation in the Deep South. As far as I know, he had no previous interest in space science until he was introduced to us in connection with the Society forming a nonprofit group to advance science and human achievement. It was the profoundness of the subject and the possibilities of what we might learn that brought him from the ken of social commentary to that of exploring the universe. The questions of human life's uniqueness (or not) resonate deeply at many levels within people: in our religions (as the experience with the Dalai Lama illustrates), myths and world views, folklore, and stories throughout all human history. The human interest can be profound—imbedding it in philosophies such as Aristotle's "prime mover" and the influential thirteenth-century philosopher and theologian Thomas Aquinas's "perfectness" as an attribute of God—or ridiculous, as in a tabloid report of alien abductions, or in popular culture ideas of "ancient astronauts."[5] I'll return to the ridiculous later. Religion is largely silent regarding questions of extraterrestrial life, and where there is commentary it is almost exclusively focused on questions of ETI. Simple life on other worlds is more or less a subject of science (exobiology as it was called, or astrobiology as is now more common), not of religion. That is not surprising since religion generally concerns those human characteristics that define intelligence— our self-awareness, including awareness of our environment, and our ability to make our presence known to others. Religion is created by humans, and as it is only a human construct it is not surprising that commentary on extra-terrestrials in a religious context is implicitly about intelligent life—beings. A cursory summary of that commentary is:

- Christian: Fundamental to all Christianity is belief in the redemptive act of Christ in dealing with original sin. Thus, ETI must either not exist or exist only to be converted. Up until the mid-twentieth century Christian theology said other worlds with other beings were inconsistent with God. But since then, as it became clear that many other worlds exist, the modern view allows for the possibility of ETI, but not as Christian.[6] This may imply some future theological arguments about the need for their conversion, for example if on the other world there had been no original sin—a discussion beyond the scope of this book.
- Jewish: God is universal, including allowance for ETI, but Judaism is not—it is about man's (humankind's) relationship to God. ETI would just be another of God's creations.[7]
- Mormon: The existence of other worlds and possibilities of ETI on these worlds is explicit in Mormonism. Brigham Young even spoke of beings on the Moon. Mormon Church president Joseph Fielding Smith went beyond just beings on the Moon and said, "We have brothers and sisters on other Earths."[8]
- Islam: Islam, too, makes explicit mention of the possibility of ETI. In the Qur'an it is stated, ". . . creation of the heavens and Earth and the living creatures He has scattered among them . . ."[9] As early as the eighth century (roughly one hundred years after the founding of Islam), one of their imams wrote, "God has created thousands and thousands of worlds and thousands and thousands of humankinds."[10]
- Hindu: The cosmology and general view of space and time of Hinduism is huge—much larger than our mere 13.8-billion-year physical universe or the smaller ones of Western religions. It encompasses several states of being in and out of which humans can cycle in numerous incarnations. ETI can be one of these, in a continuum of plants, animals, vegetation,

insects, and even deities, and where in the hierarchy ETI are depends on what they are, about which there is yet no evidence.[11] So, it seems that Hindus will accept ETI, and they will accept no ETI.

- Buddhist: The plurality of habitable worlds is not just consistent with Buddhism; it is intrinsic to it! "There are thousands of suns, thousands of moons, thousands of inhabited [!] worlds of varying sorts."[12] Not just an abundance of exoplanets, but an abundance of inhabited ones. And, as indicated by the quote from the Dalai Lama above, not just inhabited planets but objectives for contact. The Buddhist cosmology, like the Hindu one, is huge, much larger than our observable universe. It also is infinite, with repeated cycles of creation and destruction, analogous to the widely accepted notion of the varying expansion of the universe, and even the multiverse (multiple universes), which is now a popular astrophysical theory. But this similarity probably can be overstated—a stretch of ancient writings and beliefs to fit very scant data. For us, however, it is the widespread acceptance and interest that make the questions of extraterrestrial life ones so many people care about.

It is pretty clear that religious views vary but overall are not very different from the wide disparity of views that exist outside religion. What concerns us here is not the religious view regarding extraterrestrials but the recognition that what is significant in almost all religious thought worldwide is the question of the uniqueness of being human and our relation to the universe and all that has been created. "When I consider thy heavens, the work of thy fingers, the moon, and the stars, which thou hast ordained; What is man, that thou art mindful of him? and the son of man, that thou visited him? For thou hast made him a little lower than the angels, and hast crowned him with glory and honour. Thou madest him to have dominion over the works of thy hands; thou hast put all things under his feet: All sheep and oxen, yea, and the beasts of the field . . ." from Psalm 8 in the Jewish and Christian bibles is a basic human tenet (obviously self-promotional) that expresses the principal purpose of religion—to explain what human life is in relation to everything else known. The Search for Extraterrestrial Intelligence deals with that relation to both what is known and what is unknown. Religion is pervasive on Earth. Whatever one thinks about extraterrestrial life, it relates to those two questions—and billions of people care about it.

One might speculate on the reference to angels in the passage from Psalm 8. Is that a reference to extraterrestrials? If so, then humankind is clearly lower. It doesn't otherwise connect with anything else in the psalm—I tend to think it is just a throwaway line tossed in to deal with the unknown.

Different from religion, but another manifestation of the popular interest of this topic, is the large body of science fiction literature and movies dealing with extraterrestrials. Most deal with some form of ETI contact—either their coming here or us going there. On the one hand, the existence of ETI is generally considered by people to be a positive thing, much more uplifting than the conclusion that we are alone in the universe, while on the other hand ETI is often (if not usually) depicted in science fiction movies and books as fearsome, bent on destroying Earth or conquering humankind. The earliest science fiction writers, H. G. Wells and Edgar Rice Burroughs, shaped popular attitudes by describing the invasion of Earth (Wells) and continuous wars on Mars—albeit with beautiful women (Burroughs). This theme of invasion and warfare with aliens is perhaps a transposition of the colonial attitudes of Western societies, which were doing just that to indigenous populations in Africa and Asia. Other examples are in movies such as *The Day the Earth Stood Still, Invasion of the Body Snatchers, Alien, Independence Day*, and even *Man of Steel* and other Superman films. A notable contrast to fearsome aliens is Steven Spielberg's E.T.[13] E.T.'s adorable personality may have helped create positive attitudes to the scientific search for ETI, although the *New York Times* and others published cautions about letting our presence be known in the universe. Many hold that view (including the late Stephen Hawking) even though the passive act of observing or listening for ETI is not letting our presence be known. Television made extraterrestrials (ETs) popular (or ETs made television popular) with such examples as *Flash Gordon, Mork and Mindy, 3rd Rock from the Sun, My Favorite Martian, Battlestar Galactica*, and of course *Star Trek*. Television broadcasts are microwave signals that also escape Earth and travel outward into the universe. If there are ETIs and they intercept these broadcasts, I wonder what they will think of *Star Trek*.

These are all entertaining—did they increase popular interest in the possibility of extraterrestrials, or did that interest, already prevalent in society, increase the market for that entertainment? Undoubtedly, the answer is both. That sort of symbiosis also exists between popular media (sensationalist) interest and the pseudoscience promotion of alien abductions, visits, ancient astronauts, and cult beliefs. It sells. Unfortunately, this high popular interest can come with a susceptibility to being fooled by UFO and alien encounter anecdotes, which can result in a dichotomy—at least for a segment of the public: skepticism about things established in science (for example, evolution, climate change, and public health measures) and acceptance of bizarre alien encounter stories—albeit, maybe just for amusement. The ideas might be farfetched, but the public interest in aliens and UFOs has the same roots as the science of astrobiology and SETI—desiring to understand our place in the universe and wondering if indeed humankind is alone. Dealing with pseudoscience is an important educational objective because it is the result of uncritical thinking—which can be amusing, but also dangerous. In the latter category are conspiracy theories about and ideological objections to climate change, vaccinations, and evolution, and misunderstandings about statistics. It takes patience and respect to deal with people who have these kinds of misunderstandings or suffer from misinformation, something that I learned from Carl Sagan, who excelled at dealing with popular explanations of real science. But it pays off, at least in my experience—people do respond well to clear and good explanations. In this regard SETI-related questions are very valuable for educating people about astronomy, physics, biology, and many other sciences.[14]

Granting that SETI is a scientific activity, we still must acknowledge that its assumption is no more valid than are those of pseudoscience proponents who assume alien visitors have come or will come to Earth. Both presume the likelihood of ETI. And scientists are no less susceptible to allowing their beliefs to influence their work. A particularly incisive example is that of Percival Lowell, who did much for the science of astronomy and whose telescopic observations of Mars were the best of his time. His interpretation of the telescope images went far beyond the actual information contained in the data. He interpreted the presence of dark markings on the planet (caused by large surface features) as evidence supporting elaborate narratives that involved a supposed ancient civilization of Martians frantically building canals to carry water to a drying planet. That is not too dissimilar from Erich von Däniken's interpretation of lines in South American deserts as evidence of ancient astronaut visitors here on Earth from other worlds. Von Däniken is generally considered a charlatan, while

Lowell remains a respected, if misguided, scientist.[15] The important point for us to recognize here is that the popular interest in extraterrestrial possibilities (life out there, or life visiting Earth) is based on a strong human desire to understand ourselves and our relation to the universe.

The high interest in and attention paid to UFO sightings is another manifestation of public interest regarding ETI as well as a willingness to suspend critical thinking in the hope of real contact. Accepting the possibility of alien visitors is not less scientific than accepting the possibility of alien broadcasters. Both rely on narratives without data for their hypothesis of how ETI would make its presence known. The skepticism regarding UFO sightings comes not only from their extraordinary hypothesis, but because they are unaccompanied by any data (let alone extraordinary data) for scientific analysis or critical interpretation using the scientific method—which requires data that can be repeatedly analyzed and verified.[16] They are anecdotes. Their popularity is enormous (and they create great stories)—millions read the tabloid accounts of supposed alien encounters, and it seems to have taken over some of the cable channels, even the supposedly credible (at times) History Channel—drawing thousands to conventions and shows depicting extraterrestrial aliens.

According to a recent (2020) CBS News poll, two-thirds of Americans think there is intelligent life on other planets. The poll was only done in the U.S., but the results are not a particularly American phenomenon. The belief is widely shared around the world. Also of note is that the poll numbers have gone up in the last decade—ten years earlier the number was only 47 percent. People are not however expecting an immediate answer—50 percent in the poll thought it would take more than one hundred years to learn of the evidence of extraterrestrial intelligence. I have already mentioned that living in a universe that is populated with others is generally considered a more positive and optimistic answer than living in a universe where we are alone. That said, being alone would certainly make us and our world special and still provide an opportunity for humans to explore and create in a universe in which life may be abundant but we are the only "intelligent" species. Nonetheless, the conventional optimistic view regarding alien life influences the results of polling about it. The point of this book is not to critique such popular attention, but only to cite it as evidence of public interest. In fact, in this book, I welcome that interest as an opportunity to contrast scientific interest in extraterrestrial life with the pseudoscience, an opportunity to convey information and promote critical thinking. I would like to harness that interest to advance the scientific exploration of the universe. UFO stories range from seeing unexplained things in the night sky to stories of alien abductions or bizarre encounters with little green men or the like. Unexplained things are common, and most of them do get or can be explained by such things as weather balloons, atmospheric conditions, or reflected lights, but some don't. Furthermore some (very few) phenomena are observed by credible and trained observers who are careful to rule out the simpler explanations—people such as aviation, military, and law enforcement personnel and occasionally (rarely) others with scientific training. A credible observation of an unidentified object is still unlikely to be an alien encounter, but it should not be dismissed out of hand. In 2021, news reports confirmed the existence of a U.S. military program on "Advanced Aerospace Threat Identification" and revealed videos in which trained pilots marvel over unidentified objects apparently defying the limits of known technology. It is a legitimate function of government to conduct an inquiry in the public interest, although its rationale should be solid and the cost must be weighed against that of other public interests. When I served briefly as Congressional Science Fellow on a U.S. Senate Committee staff, I supported the idea of studying credible observations, but not the acceptance of the alien visitor hypothesis until real scientific proof (as a result of testing and measurements) was available. Occam's razor should be applied—that is the scientific principle that tells us that in the absence of any contradictions one should assume the simplest possible explanation. To its credit, that is just what the U.S. government did when it finally released a report on UFOs—cited in the next chapter where I consider the existence (or not) of alien intelligence. Later on in this book (chapter 6), when discussing interstellar flight, one particular UFO story promulgated by a distinguished Harvard University astronomer is discussed. He misunderstands Occam's razor and concludes that an asteroid-like celestial object observed to be travelling through our solar system on an interstellar trajectory is a craft sent here by an alien civilization.

Starting with the invention of the telescope (in the seventeenth century) and particularly the radio telescope (in the first half of the twentieth century), the consideration of ETI moved into the realm of science. It is still speculative just like pseudoscience, but the scientific method itself is more rigorous, based on facts without claims to knowledge that cannot be verified. The common misconception of pseudoscience advocates is that all theories or all ideas are equal, and we should be allowed to pick our favorite as opposed to testing them by observation, experimentation, measurement, and other tools of science.[17] As noted earlier, "extraordinary claims require extraordinary evidence." I can take a friend's word that we have a visitor from France in the neighborhood—that is not an extraordinary claim. I can't take his word that we have an extraterrestrial visitor; that extraordinary claim would require a whole lot of extraordinary evidence and analysis. We have none.

But as this brief review shows, there is still a high expectation that ETI will be found. With billions of stars and planets, and billions of galaxies, it seems a little arrogant and narrow-minded to assume we are the only intelligent species. Such thinking led astronomers to begin a search for extraterrestrial intelligence around stars in our Galaxy by listening (with radio telescopes) and looking (with optical telescopes) for signals that might indicate the presence of a technological species transmitting to space. SETI has been going on for more than a half century, so far with no positive result. In the next chapter we will find out that is not likely to change.

2

INTELLIGENCE

There is no evidence that intelligence helps species to survive.
There is some evidence that it does not.

W E   H A V E   O N E and only one example of an intelligent species. Us—
humans. By intelligence I mean life that is capable of understanding
its surroundings and forms a civilization with the ability to communicate that understanding. That is, an intelligent species has both science and
technology.[1] The science is that of understanding its environment, both
the physical and biological worlds. The technology can be many things,
but by specifically citing communications as the necessary technology, I
make it relevant to us. If they have technology but can't communicate, then
we probably will have no way of knowing about them. Unless, of course,
they just show up—but that is a form of communication. This communication criterion is only for this chapter—in future chapters I will admit
the possibilities of other technologies that can indicate intelligence—
building things that improve life, for example. The reason for focusing on
communications first is that the search for extraterrestrial life started with
the search for extraterrestrial intelligence. This was because communication was the first technology we had to search for extraterrestrial life with,
the only technology we had before the Space Age. Radio was invented
decades before space travel, and we learned to transmit radio signals long
before we learned to transmit optical (or any other wavelength) signals
or learned to send rockets off the planet. Earthly experience tells us that

radio transmission is lower cost and technologically easier than light trans-mission[2] and that both are lower cost and technologically simpler than sending rockets and spacecraft to other worlds. But that is our experience, and one should not assume the same for extraterrestrials about which we know nothing. In this book I cite both the lack of any evidence of ETI and conclusions from the years of observations that make its existence unlikely, and with that comes the conclusion to not spend much time or money looking for any technological signatures.

You might say specifying communications is just part of a larger bias—specifying technology as a defining characteristic of intelligence. There are some good arguments in favor of considering other species on Earth, including chimpanzees, dolphins, whales, and elephants, as intelligent. Aspects of their behavior appear intelligent. Of them, only the chimpanzee seems to have developed technology—if you grant that primitive tools to aid in their tasks is a technology. But that distinction may well be (likely) due to physical characteristics of chimps—they have hands and fingers. No matter, none of the species has ever made anything of themselves in the sense of forming a civilization or advancing their capabilities for build-ing things. Another reason to not consider them intelligent, even while admiring their smartness, is our continued inability to communicate with them. We could blame our own intelligence for that failure, but (1) that would be unsatisfying, (2) we're the ones writing history and science, and (3) our means of communication are rapidly evolving and adapting, and theirs—not so much. Those first two points are rather flippant, but the third is fundamental to the notion of species interaction—terrestrial or extraterrestrial.

Logically we might have wished to search for and discover primitive life first and then build up the capability to search for intelligence. But, as noted, the technologies happened in reverse—searching for extraterres-trial intelligence is largely a low-cost, Earth-bound activity, while searching for simple life on other worlds requires first getting there—that is, mas-tering space travel. Searching for extraterrestrial intelligence also can be done by doing nothing—waiting for it to reveal itself from other worlds or by coming to this one. We've been doing that passively for centuries, and actively for decades. Despite all that, nothing has been found. We have no evidence, let alone proof (or even hints) of extraterrestrial life. We don't hear it, don't see it, don't smell it, can't touch it, or taste it. Neither can we find anything that may be a remnant or a product of it. The universe could be highly populated with intelligent species, or we might be unique. Any conclusion is conjecture.

In the past half century, we have conducted several all-sky radio surveys in both the Northern and Southern Hemispheres. All-sky comes at the expense of sensitivity—that is, we can't focus as much when we search the whole sky. We make up for that in targeted searches, those that focus on a few areas better with greater sensitivity. The best all-sky surveys have been:

- META and BETA (megachannel and billion channel extraterrestrial array, respectively) led by Harvard University and principally funded by The Planetary Society.[3] It conducted searches from Massachusetts and from Argentina.
- A Jet Propulsion Laboratory (JPL) project only briefly funded by NASA in 1992 but terminated in 1993 when Congress rejected it. Its search used the Goldstone Antenna of NASA's Deep Space Network in California.
- SERENDIP, a privately funded project of the University of California, Berkeley. It was conducted using the Arecibo telescope.

The notable targeted searches are:

- A NASA program at the Ames Research Center using the Arecibo radio telescope. It was a companion to the JPL all-sky one-year search mentioned above. It was cancelled also by the aforementioned action by the U.S. Congress.
- Project Phoenix—led by the SETI Institute and formed as a privately funded activity to resurrect the cancelled NASA targeted-search program. It conducted searches from Arecibo, Green Bank, and from the Southern Hemisphere in Australia.

More recently, the SETI Institute has initiated operation of the Allen Telescope Array in California. Separately, Breakthrough Initiatives has initiated Breakthrough Listen in California, Australia, and China (at the new largest radio telescope in the world). The Allen Telescope Array and Breakthrough Listen are enabled by billionaires whose motivation is the discovery of life beyond Earth. That motivation and fact that it is privately funded illustrates the almost contradictory aspects at the basis of

SETI—extraordinary public interest but low political standing. In addition, the SETI Institute, The Planetary Society, and Breakthrough Initiatives are now supporting (with more private funds) searches using optical
telescopes to look for laser or other light signals from putative alien civilizations. Compared to radio signals that broadly spread out through space,
light can be better focused (targeted), and therefore light signals can provide higher power and stronger signals for communications—if only we
knew where they came from, or where to direct them. It is possible that,
in the future, ways will be invented to use other wavelengths of the electromagnetic spectrum for long-range, coherent communication—gamma
rays, X-rays, and neutrinos, for example—but such is not possible now.

As noted, thus far none of the searches have revealed any evidence of
extraterrestrial life. Searching is never complete, and we can never prove an
absolute negative, but we can conclude that no civilization in the Galaxy
is now trying to get our attention, either with powerful (explained below)
radio or optical signals or by visiting us. We can even put limits on the
power of hypothetical civilizations who might be "leaking" their power
or signals into space. Perhaps everyone is merely listening (or looking),
or perhaps we are the most advanced civilization in the Galaxy and many
more will follow us. Or perhaps "they" are using neutrino broadcasts or
unknown "q-rays," waiting for us to advance to their level. But these con-
jectures or any other hoped-for explanation are still only in the domains
of science fiction or wishful thinking, with nothing to back them up. What
I mean by "powerful" is best explained by the notion of Kardashev Civilizations—a concept introduced by the Russian physicist and radio astronomer Nikolai Kardashev in 1964 to characterize supposed extraterrestrial
civilizations.[4] He defined three types, now known as Kardashev I, II, and
III. Type I uses the equivalent of all the power available on a terrestrial
planet—about ten million billion watts. (Total usage on Earth right now
is about ten thousand times less than this.) A type II civilization harnesses
all the power of its parent star—about ten billion times the power of a
terrestrial civilization. How could it do this? The physicist Freeman Dyson
imagined a megastructure sphere capturing all that power built by a very
advanced civilization.[5] A type III civilization captures all the energy in its
galaxy (!), another factor of a hundred billion coming from the hundred
billion estimated stars in a galaxy. Mikael Flodin, after carrying out a rather
detailed mathematical analysis involving a statistical study of all possible

transmitter signals and SETI results, has concluded that, if there were any type II or type III Kardashev civilizations, they would have been detected by now from already existing SETI observations.[6] Flodin's analysis has not received much attention in the astronomical community—it should. Perhaps it is flawed, but if it is not, its results are profound for SETI. The analysis went on to show that if a signal were detected from a type I civilization, then it would likely be the case that hundreds or thousands of such signals exist and we somehow have missed them. There is another unfortunate conclusion that comes from this analysis. There is no good way to theorize where, what, when, or even how an extraterrestrial signal might be sent. There is no theory for predicting either the location or the nature of an extraterrestrial intelligence signal. We can only search blindly.[7] It still may be worth doing, so long as it doesn't cost too much or prevent us from doing other science. After all, analyses like Flodin's and other negative results are statistical and there is always the small possibility of detecting an extraterrestrial civilization's signal. But it seems very small, indeed. SETI's value is also enhanced because it stimulates a lot of technological development and serendipitous discoveries in astronomy. Most serious SETI programs contribute significantly to radio astronomy. SETI science is also conducted with scientific tools, and it advances science and associated engineering even when it fails to come up with definitive answers or conclusions.

Although SETI searches are basically blind—that is, without a particular rationale for where to look or at what frequency—they use scientific tools and science data. With this scientific basis and because those involved use critical thinking and pursue extraordinary evidence, science, and engineering benefit. And, as noted above, the study of the key questions about life is of enormous educational and public benefit. This is why I had no trouble supporting SETI projects even while being skeptical about their basic assumption of ETI's existence. Another point in favor of SETI projects is that they are relatively inexpensive—much less than a space exploration mission or building massive facilities. If SETI cost a billion dollars per year, perhaps we would not have been so enthusiastic about supporting it, but at The Planetary Society it was more on the order of hundreds of thousands of dollars per year. Even at NASA, it was a relatively small few million dollars per year—although that proved too much for political attacks that cancelled their efforts. The desire to find ETI, contact ETI, or be visited by ETI is strong. Scientists should be supported in their research to understand our place in the universe and to communicate about it to the public. As had been said before (in this book and elsewhere), whether ETI exists or does not—either answer is profound.

Radio and optical astronomical searches are not the only way we look for extraterrestrial intelligence. Exoplanet observations may reveal technologies—signs of technology without communications—which may be from a biological intelligent species or an artificial intelligence; this is discussed later. I noted in the last chapter that there are also the many claims of observing unidentified flying objects—or as a new government report now calls them, unidentified aerial phenomena (UAP). The evidence is only anecdotal—nothing has been physically found that can be linked to ETI. But the anecdotes abound, and some of the observational claims come from those who might be considered as credible observers—airplane pilots, for example.[8] These reports seem to come in waves, perhaps due to the power of suggestion, but they are so persistent that some effort has been made to responsibly compile and study them. A recent intense effort was made by the U.S. Department of Defense, which studied reports from military pilots reporting the UFOs. It resulted in a report released to the public because of popular interest expressed in the U.S. Congress. The report was comprehensive but could conclude nothing about the likelihood of UFOs (or UAPs as they now prefer) being from actual ETI. As noted, you can't prove a negative—you never prove something does not exist, no matter how unlikely it seems. The report did find that some of the UAPs could not be explained by any known means. But even unexplained, "extraordinary claims demand extraordinary evidence," and there is no such extraordinary evidence—just grainy photos taken at high speed. Whatever they are, ETI is hardly the simplest answer we can think of, so Occam's razor, as well as common sense, demands we reject the most bizarre option—that is, the possibility of ETIs flying (or causing something to be flown) into Earth's atmosphere near the U.S. coastline to be observed by a few people every year or so, only to disappear.

3

WHERE ARE THEY?—WE ARE ALONE

Empirical evidence suggests that intelligent life is, at most, rare in the universe.

POPULAR OPINION SEEMS to be in favor of there being life out there—I cited many examples of such opinion in the first chapter. In recent times, opinion has been highly influenced by astronomical discoveries of many planets, a large fraction of which are located in what is called the "habitable zone" of a star system with reasonable temperatures and with atmospheres, even some with hints of clouds containing water vapor. In addition, there are new discoveries of niches for life on Earth in extreme environments: deep in the ocean in hydrothermal vents, inside rocks away from sunlight, in permafrost, in very dry deserts and valleys, etc. In general, science keeps discovering more and more evidence diminishing anthropocentric arguments—that is, arguments that make us special in the cosmos. Those arguments started diminishing with findings of Copernicus and Galileo that Earth wasn't the center of the universe. Before their discoveries in the sixteenth and seventeenth centuries, the Ptolemaic principle dominated, in which Earth was the center of the universe and the stars and planets orbited it—enunciated by the second-century Greek mathematician and philosopher Ptolemy. Discoveries have abounded since Copernicus that have further reduced the specialness of our place in the universe: our solar system is not the center of the Galaxy (in fact, it is way out toward the edge), our Galaxy is not the center of its local cluster of galaxies, and those in turn are not the center of anything. We might think, given these billions of planets and given that the universe is about 13.8 billion years old, that there would be many opportunities in the universe for life to form, evolve, get smart, make things, and explore. That is what we humans did. If our place in the universe is nothing special, then why should we be special—there must be more examples. Despite all the searches and all the tens of thousands of years of human awareness, we have no evidence of "them" (whatever "they" are). Where are they? This is the Fermi Paradox.

The dichotomy between so many possibilities and zero evidence is indeed a paradox. Enrico Fermi, the person to whom the paradox is attributed, was no slouch. His question wasn't born of ignorance. He was one of the greatest physicists of the twentieth century, the creator of the first nuclear reactor and an important part of the famed Manhattan Project, which produced the atomic bomb. Quite the opposite of ignorant, his question was profound. During a conversation over lunch with other scientists about the possibilities of extraterrestrial life, he questioned the assumption that there was other intelligent life in the universe. I question that assumption, too. But, first, let me admit, no one can prove that ETI does not exist. We can't say now, nor will we ever be able to say, that there is definitely no intelligent life in the universe. We can't say that because you cannot prove a negative in science. No matter how much you lower the probability of existence, there is always a way to hypothesize how cleverly it is hiding, how rare it might be, or how dumb we are not to recognize it. But with all our observing throughout the universe, all our exploring throughout the solar system, and all our data gathering from planets far and wide, if we keep finding no evidence of ETI—isn't it reasonable to hypothesize that it does not exist?

The usual answer from those involved in the search for extraterrestrial intelligence is to note that the absence of proof is not the same as proof of absence and that we have only searched a very little bit of space, with very few (or still too weak) instruments and at very few wavelengths. However, as I described in the previous chapter, the SETI results to date do allow us to make some solid conclusions about the nonexistence of ETI. Perhaps, however, we just need to search harder, but that same argument could be made in favor of organizing and funding search parties to look in remoter areas of Earth for alien artifacts that might have landed, such as in our oceans. We can't prove aliens have not landed and left artifacts, and we have only observed a tiny fraction of the ocean floor. Yet few (if any) would be motivated to support a SETI program looking for artifacts at the bottoms of oceans (or anywhere else). Looking for extraterrestrial signals and looking for alien artifacts test the same hypothesis—namely, that intelligent life with technological capabilities beyond ours is prevalent in the Galaxy.

There is another possibility that intelligent life is just too short-lived. After all, the only intelligent life we know of is us, and the human species is less than a half million years old compared to the universe's 13.8 billion years or Earth's 4.5 billion years. Out of that mere half million years (0.0005 times the age of Earth), we've only been intelligent (that is, with science and technology) for a couple thousand years—1/2,000,000 of the age of Earth. The longevity of intelligence is indeed a huge uncertainty about which we have no information. From an evolutionary perspective, its recent presence in one species on Earth makes it seem rather ephemeral, maybe only an accident; from a technology perspective, perhaps even more so—two hundred years ago we didn't have electricity, and we can predict that two hundred years from now artificial intelligence (AI) will be widespread, and perhaps even dominant. It is a hot debate whether AI will satisfy the simplistic definition I have given here about being intelligent (aware of our universe and having a technological capability to communicate). The debate is about the "awareness" part, not so much about the technological capability. It is way beyond the scope of this book to analyze an AI's self-awareness, but even granting that AI is intelligent, we don't know how deep or long might be its existence. Existential threats to intelligent life seem enormous compared to existential threats to simple life (microbes, bugs, fish, reptiles). The latter has existed for billions of years, the former is threatened by nuclear war, climate change, asteroid impact, pandemics, and simple mistakes in programming the genetic code or our electronic equivalents in an AI world. Lifetimes of just hundreds of thousands of years (or even just tens of thousands of years) for intelligent species may be the rule.

That may seem more gloomy that I really want to convey. Gloominess is not appropriate because (1) I may be wrong, and (2) even if I'm not wrong, it doesn't make what WE do with our lives any less significant or important. The absence of ETI still leaves us with the potential for a lot of more primitive and simple life to explore. And even if the human species is limited to only another hundred thousand years—that's huge.[1] I will discuss in later chapters that there is an abundance of worlds, likely with an abundance of life to explore and understand. In doing so, we are exploring and helping to understand ourselves—taking care of our planet and finding our place in the universe. That is pretty exciting and certainly positive. For me, the absence of ETI is not a negative result.

The argument is given more apparent rigor when it is expressed in an equation—the famous Drake equation (named for astronomer Frank Drake, who at the time he invented the equation was the director of the world's largest radio telescope, Arecibo Observatory in Puerto Rico). It calculates the number of civilizations in the Milky Way galaxy whose electromagnetic emissions are detectable as the product of the following individual factors:

R_* = the rate of formation of stars suitable for the development of intelligent life (number per year)[2]

f_p = the fraction of those stars with planetary systems

n_e = the number of planets, per solar system, with an environment suitable for life

f_l = the fraction of suitable planets on which life actually appears

f_i = the fraction of life-bearing planets on which intelligent life emerges

f_c = the fraction of civilizations that develop a technology that produces detectable signs of their existence

L = the average length of time such civilizations produce such signs (in years)

The first three terms, R_*, f_p, and n_e, are clearly in the domain of astronomers. The next two, f_l and f_i, are in the domain of biologists. The last two, f_c and L, might be said to be in the domain of anthropologists. I will discuss these factors as I review the latest discoveries in astronomy about potentially habitable planets, the history of life on Earth as it evolved from its single-cell beginning into the complexity of us and other large mammals, and then the conditions that might influence social scientists to predict the lifetime of our species. But before I do that here is an overview of those factors with a range of estimates for each of them. After this overview (in table 1 below), I'll make an educated guess for each one of the parameters. Putting my guesses on paper takes great courage—because my guesses will certainly be critiqued by experts who know more about each of those parameters in the equation than do I. But my guesses are just that and incapable of being proved wrong anytime soon.

Let's estimate them:

TABLE 1 Estimates of Factors in the Drake Equation

TERM	ESTIMATE[a]	GUESS	COMMENTS
R_*	1–10 per year	3	For stable, non-massive stars
f_p	0.75–1	0.8	New discoveries indicate almost all stars have planets
n_e	0.1–0.5	0.2	More habitable planets are discovered as techniques improve
f_l	0.2–1	0.9	Based on Earth's history, with quick formation of life
f_i	0.0001–1	0.05	Intelligence could be an accident of evolution or inevitable—Earth has had it for <0.01% of its history
f_c	0.1–0.5	0.1	On Earth, we have a few "intelligent" species, but only one developed technology
L	200–500	500	Numerous existential threats: nuclear war, climate change, asteroid impact, pandemic, genetic engineering mistakes, extreme reduction in biodiversity, artificial intelligence

[a] The estimates are from the scientific literature, the guesses are personal. The reader is welcome to make their own guesses and multiply them together for their answer.

As stated, the science behind these factors is discussed later as I consider the newest discoveries about exoplanets and about the evolution of life on Earth. For now, I just summarize what is known. Notable in these factors is the big increase in planetary discoveries, tempered by the difficulties and unpredictability of evolutionary paths. The guesses are my personal educated guesses, but they are not based on any special expertise about any one of the numbers.

The product of these guesses is 1.08 civilizations in the Milky Way galaxy.[3] That is, we are probably it! ETI is a long shot, likely not present in the Galaxy. The calculation is consistent with the zero evidence we have found from SETI or hypothesized alien visits so far. However, if we take only the first four terms in the Drake equation, their product is 0.43. Multiplying that by the time that Earth life has been around (about four billion years) and assuming that it is the same for any planet, the Drake equation indicates there may be almost two billion planets with extraterrestrial life! Simple life. Those last three terms in the Drake equation all deal with "intelligent" life (as I defined it earlier—the capacity to understand our environment and to create technology to communicate). The likelihood of finding intelligent life in the universe is very small, but the likelihood of finding life in the universe is very big.

Life may be ubiquitous, but we may well be unique. While astronomers have all those large numbers that make the case for the number of worlds that are candidates for life, biologists tend to focus on the uncertainty of evolution and the difficulty of getting from simple organisms to complex beings. The astronomer versus biologist debate was poignantly illustrated by a debate between astronomer Carl Sagan and biologist Ernst Mayr, published in the *Planetary Report* in 1996. Despite all that has occurred since then—several comprehensive SETI observation programs with both radio and optical searches, thousands of exoplanet discoveries from both space-based and ground-based observatories, the sequencing of the human genetic code, measurement of the cosmic background radiation and other advances in cosmology, the discovery of additional fossil evidence of human ancestry, new data and understanding of planetary processes including climate change, etc.—there is little new information affecting their arguments about the probability or improbability of intelligence. Both agree about the high probability of extraterrestrial life and strongly disagree about the probability of such life adapting toward intelligence. Rather than attempt to represent their arguments, I reprint their discussion in appendix B.

A different argument is contained in the Cosmic Zoo hypothesis.[4] It explains the lack of ETI contact by concluding we are not worth contacting. The theory supposes that evolution to complex life is more or less inevitable (or at least common), leaving open the question of whether complex life develops technology. This conclusion comes from a comprehensive examination of evolutionary pathways for Earth life, which purports they all more or less lead to complex life. It is a well-reasoned argument that has no shred of supporting data and in fact is contradicted by two huge observations: the aforementioned lack of complex life for half of Earth's evolutionary history, and all those random, internal, or celestial interventions from geology and astronomy in the biology of evolution. Things such as orbit variations (which cause climate changes), asteroid impact, volcanism, magnetic field anomalies, etc. Again, we have one and only one example of complex life—that on Earth. And, for now, it is inconsistent with our being in a cosmic zoo. As we look around those planets that we have seen up close, we can believe in simple microorganism life (say, on Mars or on Jupiter's moon Europa), but we are hard-pressed to imagine complex life, animals, giant forests, etc., on these places. The authors of the book *The Cosmic Zoo: Complex Life on Many Worlds* do acknowledge two weaknesses—they say the origin of life is improbable, and they have no way to guess how technology is a likely or possible outcome of complex life. That first weakness is not likely—the origin of life may not be improbable. It happened here on Earth quickly, and the precursor conditions for life are everywhere in the universe. But common or not, it does not affect their argument. Getting technology from complexity is indeed hard (it happened only once here among all of our species on Earth), but getting complexity from simple cells is also hard (it took Earth more than two billion years). In any case, the Cosmic Zoo hypothesis for Earth's life is unlikely.

Let's return to that huge difference between the time scales associated with the start of life on Earth and its evolution to intelligence. The former number was 3.5 to 3.8 billion years ago, a "mere" 0.75 to 1 billion years after Earth formed. Is that just a happenstance, or is that typical of planets everywhere? I noted earlier that intelligence (including the creation of technology) has only been around for $1/2,000,000$ of that time—just the last couple thousand years. Life has been on Earth for about 85 percent of its existence; intelligence has been on Earth for about 0.0005 percent of that time. Optimists might want to argue that intelligence is only at its beginning, and after a million years or so those numbers will drastically change, perhaps with intelligence occupying a greater portion of Earth's history. But that is a lot of optimism, especially in the absence of any other evidence about intelligence in the universe. The upper limit for Earth's survival is only about two-to-four billion years (when the Sun starts to die), and we can probably expect that intelligence will die out a lot sooner than that—whether by existential forces (some natural, some human-caused) such as climate change, asteroid impact, nuclear war, pandemic, or by the evolution of other species or technologies to replace intelligence with something else. The fact that we have such a list of possible extinction directions is almost proof that something like that will happen.

There is a profound question of definition here: what do I mean by species survival? In strict scientific terms it is well defined—*Homo sapiens* is our species. We *Homo sapiens* evolved from and with other species, but all those species of our more general *Homo* (which means "man" in Latin) who were contemporaneous with *Homo sapiens* ("wise man") died out: for example, the Neanderthals. There is debate about the future of human evolution: scientists have three broad views (and many much more detailed ones). (1) Evolution is done, and we are the end of (at least) Darwinian evolution, having now reached the stage of modifying our environment and conditions for survival. (2) Human evolution is actually increasing in speed with genetic modification already being noticed over time periods of a few hundred years,[5] which will lead to a new human species. (3) Our interaction with robotics and AI will produce a symbiosis of human and machine with qualities of a new species. That last idea is confusing—do we call the symbiosis with machines a new species?[6] Inventor and futurist Ray Kurzweil takes this argument much further, saying that in just a few decades from now our genetic evolution will be a mix of genes and silicon chips—leading to a post-human species derived from AI, robotic technology, and human genetics. He dubs it the Singularity.[7] That may be a stretch (I certainly hope so). Such ideas are explored even more deeply in the trilogy by Yuval Noah Harari, *Sapiens: A Brief History of Humankind* (Signal, 2015), *Homo Deus: A Brief History of Tomorrow* (Random House, 2016), and *21 Lessons for the 21st Century* (Spiegel and Grau, 2018). In *Homo Deus*, he writes, "A non-organic artificial intelligence . . . will find it far easier to colonize alien planets. The replacement of organic life by inorganic beings sows the seeds of a future galactic empire, ruled by the likes of Mr. Data rather than Captain Kirk."[8] In science fiction, robotic intelligent beings are generally considered life of a different form, but how biologists and life scientists on Earth will view this is not yet clear. It might simply come down to Descartes's maxim: "I think therefore I am."[9] In that case, future thinking machines will be life. This speculation is way outside the scope of this book, which is all right because, as I consider extraterrestrial life, I am only concerned about its existence, not its evolution. If, perchance, we happened upon a machine, nonbiological civilization of no living beings but only their robotic descendants (or their self-replicating von Neumann probes[10]), we would certainly regard that as contact with extraterrestrial intelligence. However, the very fact that we have not seen any sign of ETI, organic or inorganic, is a hint of its unlikely existence. I know well that the absence of evidence is not evidence of absence. However, here we must combine the absence of evidence with what we have learned about the evolution of planets, life, and technology. If ETI ever existed anywhere (as I have defined it throughout this chapter—as a civilization capable of communicating its existence), then again I ask "Where are they?" Surely, given the thirteen-billion-year history of the universe, if there were such, there would be some hint. Flodin's thesis cited in chapter 2 quantifies this conclusion. It does not matter if "they" are biological or not. In chapter 8, I will propose that the future of human exploration of exoplanets will be done by robots, AI, and virtual reality—hopefully not by a species that replaces us, but by us, ourselves. The conclusions in this chapter suggest we are the only species in the universe (or at least the Galaxy[11]) capable of doing that.

Most species survive for far less than a billion years—especially complex ones. I also note in chapter 4, when discussing the notion of "Rare Earths," that evolution from simple life to intelligence is laden with lots of random, apparently lucky events. Humans may have emerged on Earth by random luck, and the cards continue to be stacked against us to avoid extinction. Of the millions of species on primitive Earth and all their evolved lineages, only one made it to intelligence, that is, only one is capable of making its presence known in the universe. We still debate, without knowing the answer, whether the runners-up in the race to evolve intelligence are even intelligent: for example, dolphins and chimpanzees. They may be smart, but they do not have the capability of making their presence known to the universe. And we really are unsure of how self-aware they are in relation to their environment. There is absolutely no reason (except for faith or wishful thinking) to expect that the probability that life evolves to intelligence is high. It is sometimes argued that the reason only one out of Earth's millions and millions of evolutionary pathways made it to intelligence is because humans became dominant (and wiped out or at least replaced competitors). But that doesn't hold up—dolphins have been around for fifty million years, a hundred times longer than humans. One can imagine many different evolutionary paths to intelligence and/or, particularly, to a species capable of broadcasting its existence. Indeed, the science fiction literature is full of examples, but we have no working model of an intelligent species that isn't us. This could all be upended with a discovery tomorrow—that is the lure of SETI. But until such a discovery is made, or there is even a hint of the possibility, we should assume the opposite. There is another dichotomy to ponder: the evolution of intelligence from simple life took a long, long time—billions of years. The evolution from intelligence to artificial intelligence may happen much more quickly—just thousands of years! Indeed, the time scale to move from technology capability to AI may only be hundreds of years. In a book I wrote several years ago, I argued that human spaceflight beyond Mars will only be done without humans—that is, by robots with AI and humans joining in through virtual reality using robotically gathered data.[12] This idea of future human exploration of Mars being done by humans still on Earth is exciting—it will involve billions of people for thousands of years with a planet whose surface area is equal to that of Earth's. In chapter 9, this idea is extended from Mars to thousands of exoplanets, and potentially habitable ones.

It is not a stretch to speculate that the human species may not last hundreds of thousands of years, let alone millions. Of course, we cannot say whether the posthuman species will even fit our definition of intelligence: self-aware and capable of making itself known. It is possible that many intelligent species with technological capabilities formed in the universe, only to have died out or been replaced rather quickly (in thousands of years, not millions or billions) by noncommunicating species or noncommunicating things. This is actually consistent with the thesis cited in the previous chapter that concluded from all the SETI observations that Type II and Type III civilizations don't exist, and that if no Type I civilization is discovered then there are probably none at all—since even one would mean many. This is what makes SETI and the exploration for extraterrestrial life so interesting, even without data—it makes us ponder the really big questions. Like this one: what is our fate?

4

AN ABUNDANCE OF WORLDS—AN ABUNDANCE OF LIFE

The universe may have no other intelligent life, but it is
teeming with simpler life—on billions of worlds.

NOW DISTINGUISH BETWEEN the search for life in the universe and
the search for intelligent life (that is, for ETI). Simple life may be abundant, but complex life seems to be rare. The step from simple life to intelligence is enormous—indeed, just getting to complexity (a precursor to
intelligence) seems to be a huge step. It took two billion years and a few
lucky circumstances for it to happen here on Earth—the only example we
have of it happening. Current studies suggest that life formed quickly after
Earth itself was formed. Earth formed in the solar system about four and a
half billion years ago, after the formation of the Sun. Rotating clouds of gas
gradually clumped together until a central clump became large enough to
start a nuclear reaction, burning the primordial hydrogen into helium and
forming the Sun. Other clumps gradually grew larger and began to form
the planets and our solar system. Remarkably, on (at least) one of these
planets, Earth, life seems to have started within a few hundred million years
after that. I say remarkably considering that, at the time Earth formed, the
universe was already about nine billion years old, and yet, as soon as our
planet formed, despite its very hostile conditions (no free oxygen and high
heat), life began here in about one-tenth of that time. We surmise this from
evidence of the first single-celled fossils that date back to more than three
and a half billion years ago. The origin of that life is unknown—it could

have formed from the chemistry and elements in the proto-Earth, or from conditions in those first few hundred million years after formation, or maybe even it was deposited here from elsewhere in space as the planets and other bodies of the solar system formed.[1]

Not only don't we know the origin of life on Earth—we can't even clearly define what life is. The boundary between life and nonlife, or between chemistry and biology, seems indistinct. A working definition of life was given in a 1992 NASA meeting connected with the initiation of a new program for SETI: "Life is a self-sustained chemical system capable of undergoing Darwinian evolution."[2] That is, of reproduction, change, and adaptation. But some chemical processes arguably have these character-istics and, even then, there are huge disagreements about, for example, whether a virus is alive or not. I think this ill-defined boundary between life and not-life is probably at the heart of the explanation for why life seem-ingly originated quickly on Earth and why we continue to be so hopeful of finding life on other worlds and in places that seem to have the right conditions for it, even without definitive proof.

But, after this quick start to life, not much happened. Life on Earth remained single-celled and simple for the next nearly three billion years! That is, life was simple for more than half of Earth's history, and we have lit-tle idea of what caused complexity—why evolution took off. Life remained single-celled until at least one billion years ago. The oldest evidence of multicell life is from about nine hundred million years ago. One hundred million years later those multicelled microbes finally picked up the pace of evolution—a pace that led to us (intelligence). Primates first appeared fifty-five million years ago (ten million years after the huge comet, or aster-oid, impact that cleared Earth of dinosaurs and many other species so that mammals could begin to take hold). The sudden takeoff of complexity, for no apparent reason, and the asteroid impact that enabled mammals to succeed make us wonder if evolution from simple life to intelligence was an accident—not easily repeated. Consider the timeline of evolutionary highlights from the formation of Earth to the appearance of human intel-ligence and technology:

- 4.5 billion years ago (Bya)[3]—Earth forms
- 3.8 Bya—life appears
- 2.5 Bya—oxygen in the atmosphere, toxic—most life is destroyed

- 0.5 Bya—the Cambrian explosion (see below)—finally life gets going with some fish
- 0.09 Bya—mammals start appearing
- 0.065 Bya—Chicxulub asteroid impact—dinosaurs and lots of other species die
- 0.006 Bya—first primates appear
- 0.0002 Bya—first humans appear[4]
- 0.0001 Bya—first writing appears
- 0.00000005 Bya (500 years ago)—physics and astronomy make us aware of the universe
- 0.00000002 Bya (200 years ago)—electricity, radio, and technology appear

Figure 1 depicts this history graphically. All human history is contained at the top line of nearly zero width. The list above ends two hundred years ago with the advent of technology that can broadcast our existence extra-terrestrially. Since that time, humankind has created the potential for self-destruction through weaponry, climate change, pandemics, genetic engineering, and the creation of AI.

As I said, we don't know whether the evolution from simple to complex life was inevitable or an accident. This timeline has a few unpredictable features—the rapid oxidation 2.5 Bya, the Cambrian explosion 0.5 Bya, the huge asteroid (or comet) impact and extinction event of 0.065 Bya (not to mention others, including the even bigger Capitanian event 0.026 Bya)—that lend credence to the idea that the evolution of mammals was a cosmic accident. If so, perhaps the Earth experience is rare. The idea of "Rare Earths" was developed in a book with that title by Peter Ward and Don Brownlee.[5] They examined geological conditions of Earth and noted how special, and possibly rare, were those that were conducive to the evolution of life. For example, take the Great Oxygenation Event. Oxygen was rare on Earth up until about two and a half to three billion years ago, but then for some reason it was produced abundantly from microbes in the ocean—it is still not well understood why. Or take plate tectonics: Earth has it, Mars and Venus don't. In fact, Earth is the only planet we know of with plate tectonics. This seems to be crucial for the rise of life—the action of plate movements on Earth alters conditions on the surface and in the atmosphere a lot, creating quakes, volcanoes, and atmospheric and

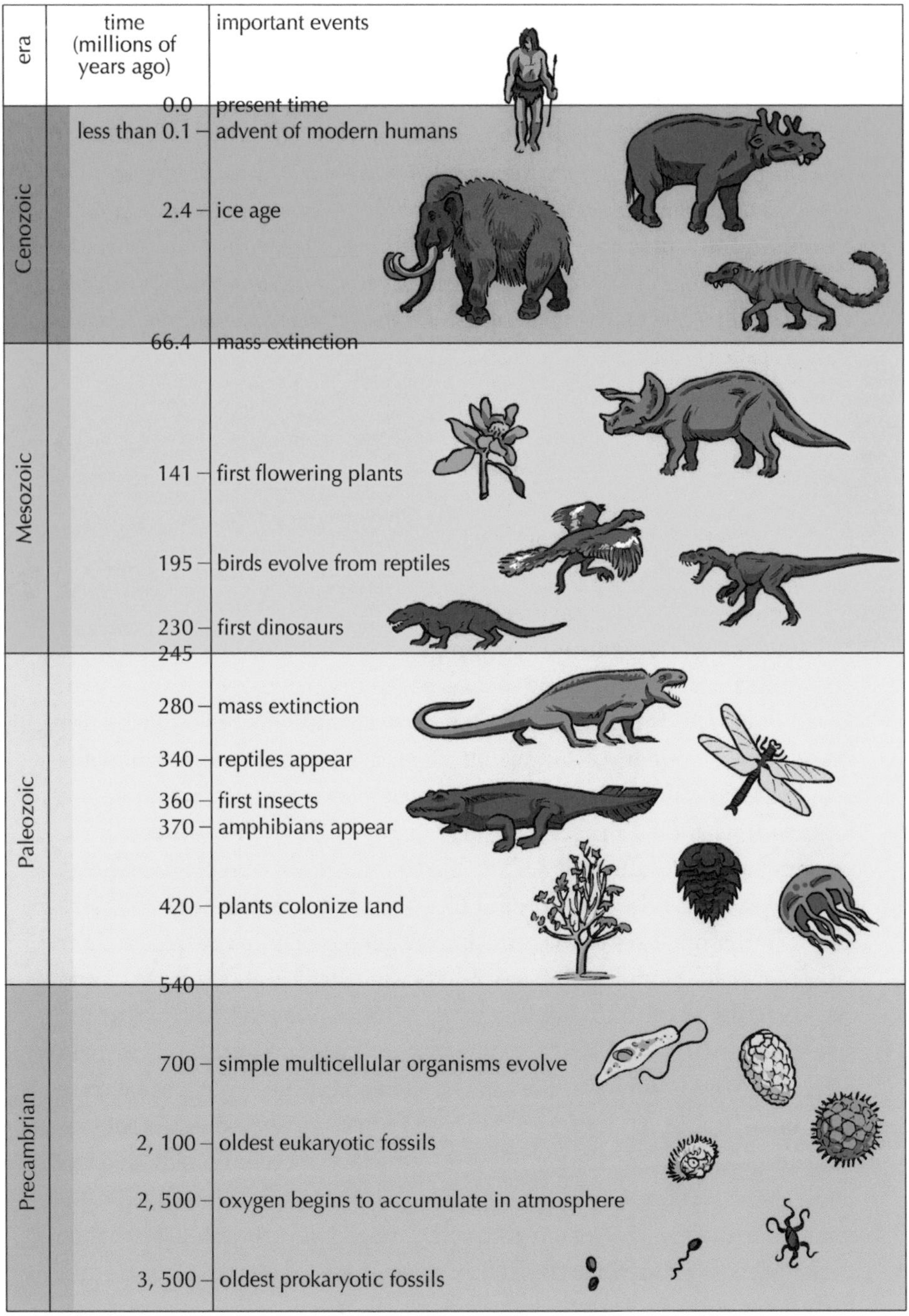

era	time (millions of years ago)	important events
	0.0	present time
Cenozoic	less than 0.1	advent of modern humans
	2.4	ice age
	66.4	mass extinction
Mesozoic	141	first flowering plants
	195	birds evolve from reptiles
	230	first dinosaurs
	245	
Paleozoic	280	mass extinction
	340	reptiles appear
	360	first insects
	370	amphibians appear
	420	plants colonize land
	540	
Precambrian	700	simple multicellular organisms evolve
	2, 100	oldest eukaryotic fossils
	2, 500	oxygen begins to accumulate in atmosphere
	3, 500	oldest prokaryotic fossils

FIGURE 1 History of life on Earth. Illustration by Siyavula (CC BY 3.0).

climate changes. This enhances the evolution of different species, which adapt to changes and improve their survivability—the basic principle of Darwinian evolution. Indeed, the changes induced in East Africa changed its lush forests to arid savannahs, which in turn led to the evolution of intelligence in *Homo sapiens* separately from that in the other hominids (such as chimpanzees) that did not so evolve.[6] Another example special condition is the presence of Earth's magnetic field—without it we would have no ionosphere to protect us from extreme ultraviolet radiation that would sterilize the planet (like Mars). Our location in the habitable zone combines with other factors such as the spin of Earth creating day-night cycles, the tilt of the axis creating seasons, and even the presence of a giant planet (Jupiter) on an orbit far away that perturbs incoming comets just so so that the water accumulation on Earth was itself just right. This lucky combination of celestial behavior made our place in the habitable zone actually habitable. In addition, we have random but crucial huge events such as the freezing of the planet ("Snowball Earth"), the Cambrian explosion,[7] and several massive extinction events from celestial impacts (such as that aforementioned wiping out of the dinosaurs) or volcanic eruptions that killed off many species of life to open the way for the evolution of others. There is no discernable cause-and-effect relationship for these events—nor any assurance that they are anything more than lucky accidents of random evolution on the planet. Putting them together and looking at what we don't see on other worlds, Ward and Brownlee make a convincing case for the rarity of Earthlike conditions favorable to complex life. They also cite the apparent fragility of complex life. If the planet were a little hotter or closer to the Sun (like Venus), or a little colder and drier (like Mars), life would be impossible. The so-called habitable zone around stars is rather small, and that fragility is compounded by the fact that the star itself (such as our Sun) must be just so—not too burnt out, not too active, etc. This is not to say life can't happen elsewhere—but it seems rare and unlikely at best. Contrast that fragility of complex life to the robustness and resiliency of simple life—found everywhere on Earth, even in hydrothermal vents without sunlight, in the insides of rocks, buried deep inside Earth, or in the frozen Antarctic. Indeed, life persisted even after the huge mass extinctions that followed asteroid (or comet) impacts and early giant volcanic eruptions that wiped out many, or in one case most, but not all (!), species.

As far as we know or can even theorize, life can only exist on planets. Sure, we can imagine microbes travelling individually or in small groups in space, say in a rock or meteoroid, but most likely they came from a planet and are temporary travellers between worlds. I noted earlier the recent spate of discoveries of exoplanets since the mid-1990s. There are an astronomically huge number of stars, and now it seems that most of them have planets. Thousands of exoplanets have been discovered with a wide range of characteristics. Billions or even trillions probably exist. There are hundreds of billions of stars in our Galaxy, and billions of galaxies in the universe, perhaps hundreds of billions—just in the observable universe.[8] Multiplying those two numbers we estimate there are perhaps 10^{23} stars (one hundred billion trillion), about one hundred thousand times the number of grains of sand on Earth, which happens to be about the same number of seconds since the beginning of the universe!

As of mid-2021, over four thousand exoplanets have been discovered—basically all in the past fifteen years. The number is steadily increasing. Almost all of these are either gas giants or large Neptune-sized solid bodies. Fewer than 150 are terrestrial, so-called "Earthlike." That disparity in the distribution is not surprising—larger planets are easier to discover because either they have more mass (causing a larger gravitational effect on their parent star), or they are bigger (easier to directly detect with telescopes). Due to other observational biases, we tend to preferentially detect planets around dimmer stars (less starlight interference) and planets with small periods (allowing for redundant observations of their orbits). Based on our understanding of our own solar system and its formation, the population of terrestrial planets is probably significantly underrepresented. Almost all the exoplanets that have been observed are in different (distinct) star systems, although several instances of multiple planets in a single star system have also been found. There is growing evidence that most stars have at least a few planets. That means billions and billions of planets just in our Galaxy, and billions and billions of times that in the universe.[9] Many of these stars are in different stages of evolution—bigger, smaller, hotter, cooler, older, and younger than our Sun. We don't know enough to rule out candidate habitable star systems yet—although planets around those stars that are a lot hotter and bigger than our Sun are unlikely to be habitable, since they would be unlikely to hold water (or any solvent). Similarly, habitable planets around burnt-out and much cooler stars are also unlikely.

Some scientists conclude that stars slightly cooler than our Sun have ideal systems, based on the theory that molecules on their planets would be very stable—but we really don't know.[10] Even with more information about terrestrial-sized planets, we won't know until we get substantive compositional data from a wide variety of exoplanets. Restricting ourselves to Sun-like stars, the latest estimate is that there are probably about three hundred million terrestrial, Earthlike, planets in our Galaxy. That's less than about one-tenth of 1 percent of the number of stars in our Galaxy—so three hundred million is a pretty conservative estimate. Multiply that by a billion galaxies and, yes, we can conclude there is an abundance of worlds out there that we might call Earthlike. But what about an abundance of life?

Here, we move into the domain of biologists. Life formed very quickly on Earth. It's been here for 85 percent of Earth's history. It may have formed here from the planet's initial chemical constituents. Or it might have been deposited here in the early stages of our planet's evolution from some other place. We know that debris from volcanoes and impacts on Mars and from comets and asteroids have regularly landed on Earth. We don't know yet about how life can survive travelling through space, or if it is possible to "seed" planets with life, that is, by the carrying of it from one planet to another (panspermia).[11] If we ever do find life elsewhere in our solar system, for example on Mars or in the subsurface ocean of Jupiter's moon Europa, we will be most interested in the forensic analysis that determines if it is exactly the same as Earth's, meaning a common origin for life and likely transpermia, or different, meaning probably different origins. For our consideration here, it does not matter whether life originated on Earth or was seeded here from afar—we just note that it started quickly on Earth. The chemical basis of life, as we know it, is found in those elements present in planetary formation: carbon, oxygen, hydrogen, nitrogen, phosphorus, and sulfur. Those elements are there at the beginning and begin interacting chemically. Maybe those chemical reactions form life quickly on other terrestrial planets, also; we don't know. Nor do we yet know exactly how life forms from these elements—is it by random chance, is it inevitable, or is it because of a peculiar set of conditions?[12] Attempts to create life using the conditions and constituents on primitive Earth have thus far proved unsuccessful. The first, and still defining, experiment was that of Stanley Miller and Harold Urey. The Miller–Urey experiment took the chemicals that make up life and mixed them in the right proportions and added external energy, just as it happened on primitive Earth.[13] They did make amino acids, which are building blocks of life, but they did not get life. Nor have any attempts since. The theory of that experiment is that life comes from the chemical and organic molecular building blocks. There are other theories based on the more recent research into RNA and DNA. It is known that RNA can act as a catalyst (DNA can't), and the so-called "RNA world" hypothesis is that a catalytic reaction produced self-replicating RNA molecules, passing genetic material forward, eventually to DNA and life. Another theory says that metabolic networks preceded and produced RNA and DNA, mixing the chemicals and molecules under high pressures (as in the interior of primitive Earth or in hydrothermal vents) to produce a self-perpetuating network of reactions. No experiments have yet been devised to test these theories and we do not yet know the origin of life on Earth, or even for certain that it started on Earth. It is this gap in our knowledge, the missing link between the chemistry of the elements and molecules that form life and the biology that starts the processes of replication and metabolism, that prevents us from knowing whether life is likely to be common in the universe. We, of course, see chemistry everywhere on the planets and in space; observing the chemistry of habitable worlds should provide much additional information about these missing links that lead to biology and the origin of life.

We have long observed the planets of our solar system (first by naked eye, then with telescopes, and now regularly with spacecraft) and wondered about the possibilities of life on other worlds.[14] We have sent spacecraft to every planet in the solar system (and landed on the two nearest ones). We have also sent spacecraft on close flybys of many of the moons of Jupiter and Saturn, and by comets and asteroids. Now we are observing thousands of exoplanets, some of which are terrestrial with atmospheres such that we might even see elements and molecules that could form life or lead to life on them—tantalizing hints of life out there even from our observations here on Earth. We send spacecraft close to, even in and on, the planets in our solar system, but we can't reach the exoplanets (see chapter 6) and thus are limited to distant, remote observations, looking at points of light in the glare of their parent stars. Exoplanet research and observation is a big field now with lots of astronomers and other scientists involved. The techniques and observables for seeking signs of life on other worlds include spectrographic observations of the atmosphere, that is, looking at the wavelengths of the reflected light from the planet (reflected from their star) to determine the chemical composition of molecules in that atmosphere. Also, observing the orbital motion of the exoplanet to determine its period, distance, and maybe even its inclination around its parent star. This tells us if the planet is in the habitable zone around its sun. The habitable zone is based primarily on the temperature of the planet, predicted by knowledge of the type of star it orbits and its temperature. It does not take into account the greenhouse effects such as that on Venus, which is much hotter than would be predicted based on its distance from the Sun, nor internal forces heating the planet, as in the case of the moons of the outer planets, which are much hotter (internally) than would be predicted based on their distance from the Sun. There could be life on worlds outside the habitable zone, but with our Earth-centric view of things we give first attention to planets in habitable zones around other stars. Unfortunately, no telescope on Earth (or in our solar system for that matter) can reasonably be built large enough to get more than a one-pixel-sized image of an exoplanet—barely enough to tell whether the planet has clouds or not, and not enough to reveal anything about continents or oceans, and very little about its composition. The largest single-aperture optical telescope on Earth is the 10.4-meter Gran Telescopio Canarias on La Palma in the Canary Islands. If observing the nearest exoplanet of the nearest star, the smallest-sized feature it could detect (the telescope's resolution) would be 2.2 million kilometers. At this resolution, a terrestrial planet would be a dot 1/100 to 1/1000 the size of that pixel—a pale dot, at best, in any telescopic image. To get a single-pixel image you would need a telescope many tens of kilometers in diameter. And you would need it to be in space to get a clear image. Clearly, this is impractical.

In chapter 7, I will present a possible way to get a magnification of one hundred billion times of an exoplanet with a space mission in our solar system. As of late 2020, of the four thousand plus exoplanet discoveries, about twenty of them are classified as the best candidates to be potentially habitable, so-called Earthlike, even though none of them are really known to be like Earth or habitable. However, as noted earlier, the non-Earthlike, not potentially habitable gas giants and other large objects are the easiest ones to observe around other stars, and we can infer there are many smaller terrestrial-sized planets even among the four thousand plus planetary systems thus far identified. All of the twenty best candidates for potentially habitable exoplanets discovered thus far are more than one hundred light-years from Earth.[15] Within fifty light-years, there are twenty-three terrestrial-sized exoplanets in their stars' habitable zones.

With so many exoplanets (more than a hundred million in our Galaxy), we can reasonably expect that there will be more than a million potentially habitable ones. Given our earlier observation about how quickly life formed on the one habitable planet we know best, and how widely spread are the basic chemical constituents for life in the universe, we are justified in concluding that life has likely formed on many planets and is possibly ubiquitous throughout the universe. This is something to explore.

5

EXPLORING NEW WORLDS—WE ARE NOT LONELY

Exploration of potentially habitable worlds provides humankind with unlimited adventure and opportunities for discovery.

So far in this book I have cited a lot of data that lead to the conclusion that, most likely, the human species is alone in our Galaxy, if not in the universe. The combination of negative results from decades of observation and the theoretical arguments about possible communicating civilizations cited in chapters 2 and 3, along with our knowledge that on Earth we are the only species (among millions) that has built a civilization, strongly suggests that we are likely the only species in the Galaxy to be intelligent—or at least to be communicating and technological. While some may find that depressing, I don't. The natural philosopher E. O. Wilson states, about the human species, "[we are] alone and free."[1] Rather than being a negative, in his view that makes us free from being subject to others or from invoking a higher authority to regulate our destiny. It's an uplifting thought. Uniqueness also makes us special. Being alone in the universe doesn't mean we are lonely. In the previous chapter I cited the abundance of worlds with varied physical conditions and different possibilities for life. We can learn about other life, which will teach us much about our own life on Earth. In this way, we may figure out the origin of life, discover multiple evolutionary paths, and ultimately determine how or why we are special (if we really are). So many worlds, not merely to explore, but to interact with and learn from—learn about ourselves, our past, our present, and our future.[2] This is hardly a condition of loneliness.

Let's start with our own solar system with its currently known population of eight planets, more than two hundred moons, more than two thousand Kuiper belt objects and a nearly uncountable number of asteroids and comets. Of course, most of these (nearly all) are not what could be considered habitable, but still there is an abundance of worlds with the ingredients of life—that is, with water, with the key elements and molecules of oxygen, carbon, hydrogen, and nitrogen, and with sources of energy—internal and external. Astrobiologists working in planetary science today say it is reasonable to look for evidence of life on (or in) Mars and in the water worlds of the outer planets. In our solar system, the terrestrial planet Mercury has likely never been habitable. Probably neither has Venus, sometimes described as being like hell with its much higher than oven-like surface temperature of 462°C (864°F) and one hundred times higher than Earth's surface pressure. Hell-like, but not living hell. Venus was not always this way—there was a Venus before the runaway greenhouse created its carbon dioxide blanket, and who knows what it might have been like.[3] And even now, high up in the atmosphere in a cloud layer, the temperature and pressure are moderate, with even some water vapor present. Water or some other liquid there could serve as a solvent to sustain life or even allow it to form.[4] However, even the scientists speculating about Venus life admit it is a rather wild idea considering that the most abundant liquid is sulfuric acid and there is no obvious place (minerals) for molecules to rest on or to catalyze. Mars is certainly a candidate for life—most likely habitable in the past when it was warmer and wetter, and perhaps even so now, with life possibly buried in niches below the surface where there is water and ice and protection from ultraviolet and cosmic-ray radiation. Unlike cloud-shrouded Venus, we have repeatedly seen all of Mars's surface—at higher and higher resolution with more sophisticated spacecraft. We know about Mars's warmer and wetter past from the clear evidence left by rivers and lakes on its surface.

Orbiters around Mars have detected water that makes it to the surface, even today, although with the cold and thin atmosphere it sublimes quickly away (figure 2). Mars is also potentially habitable in the future—if we make it there and build shelters to protect us from high ultraviolet and cosmic-ray radiation and to contain an atmosphere. Using the indigenous resources of carbon dioxide and water, we can introduce plant life and make oxygen, and, perhaps in the far future, a self-sustaining ecosystem for life. I think that Mars is a laboratory for us, on which and with which we can study and experiment to understand how widespread extraterrestrial life might be. If life was not there—on our next-door neighbor with a warm, wet past— then it might be hard to argue for its widespread distribution in a more hostile universe. If it is or was there, despite its present toxic and hostile environment, the opposite might be said. Mars: if life can't make it there, it can't make it anywhere. Maybe.

Besides Mars and Earth, we have a plethora of solid-surface moons orbiting the outer planets with the ingredients of life, indeed, with oceans and lakes that potentially could be abodes for life: the ocean worlds (figure 3).[5] Astrobiology is a relatively new subject—it was not a field of study when I started my career in planetary science. It is still a subject without subject matter (no other biology has been discovered), but what we have learned from our exploration of the solar system in the past fifty years makes it a subject of rich interest with many possibilities. We know there is internally heated water beneath the ice crust of Europa (a moon of Jupiter) and geysers of water shooting out from Enceladus (a moon of Saturn).[6] There seems to be water on other planetary moon surfaces as well. Even Titan, way out there orbiting around Saturn (and getting about 1 percent of

FIGURE 2 A lake of water ice on Mars. Liquid water is known to be below the surface. Courtesy of the European Space Agency.

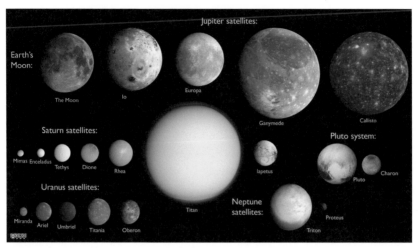

FIGURE 3 Ocean worlds in our solar system. Montage from The Planetary Society, courtesy of Francis Nimmo and Emily Lakdawalla.

the sunlight that we do here on Earth), with its atmosphere of methane and other hydrocarbons, seems to have water. Comets regularly bombard all objects in the solar system with water. We have found that life on Earth survives in the most extreme of environments—in the cold of Antarctica, in the heat of deep thermal vents in the oceans far from sunlight, inside rocks, and airborne on tiny molecules. The extreme environments of Mars and other worlds may also have niches for life. The interior ocean of Europa and possibly interior ponds on Enceladus and Titan and perhaps even Pluto likely have the elements of life—although whether they are stable enough environments for life is of course unknown. Those places are frigid, but they have internal heating in their interiors from tidal interactions caused by the gravity of the large outer planets. The space-age discoveries of interesting worlds in the outer solar system (especially the moons of Jupiter and Saturn) have either rendered obsolete or enlarged our view of planetary habitable zones—depending on your point of view. This is important when considering future exploration of exoplanets. The scientists in that field tend to talk a lot about habitable zones (i.e., the distance from the parent star), but that is because it is all they know about at this stage. We don't know anything about exoplanet moons and the stability of temperate conditions in various exoplanet environments—we have much to explore. That is where this book is heading: see chapter 8.

Since 1960, we have regularly and systematically explored the solar system—208 different missions through the end of 2020, including lunar missions (even the ones with human crews), but not including those that failed on launch.[7] Missions have been flown by the Soviet Union,[8] the United States, Europe, Japan, India, China, and most recently the United Arab Emirates. Exploring the solar system is a venture of Earth. Only one of these missions (Viking in 1975) was specifically designed to detect life, but our view of extraterrestrial life then was more naive and the experiments on it were based on preconceived and hopeful notions. It seemed then like we were waiting for that eureka moment of discovering something so recognizable we could say, "there it is." Now, we think the evidence for life might be much more subtle, so much so that some ask if we will even recognize it. What can be and is being done in addition to looking for life itself is to look for geologic and geochemical evidence of it—perhaps some fossils or other markings that could have been made by living things, and for conditions on a planet that are conducive to life—and then try to put the evidence together in a way that leads to a conclusion. But the bits and pieces of data can come from disparate and unexpected places. For example, a navigation engineer made a serendipitous discovery looking at the Voyager data, noticing a plume coming from Jupiter's small moon Io. It turned out that it wasn't just a lucky moment to witness a volcano on Io: volcanoes were erupting there all the time. That wasn't expected in the frigid temperatures out at the distance of Jupiter. Planetary scientists realized that the tidal forces inside the moon caused by its motion so close to Jupiter must be causing the interior of the moon to heat up from friction (analogous to tides causing waves on Earth in our already liquid ocean). The heat engine from the tidal force causes the constant volcanism. This has nothing to do with life—no one thinks there is life on Io. But a weaker version of that tidal force on the next moon out, Europa, causes heating underneath its surface; not enough to make volcanoes, but sufficient to melt the ice deposited there from millennia of cometary bombardment. There is so much water on Europa that it formed an ice-covered surface and has a liquid ocean underneath it. That has a lot to do with life because, the suggestion goes, where there is water and heat there is likely to be life. Scientists might have been skeptical about the possibility of an ocean 750 million kilometers from the Sun if they hadn't been conditioned by the discovery of internal heating that causes volcanoes on Io. That is how multidisciplinary science works—the gathering of disparate observations together to develop a hypothesis, one that now can be tested with future spacecraft.

Missions to Mars, Venus, Europa, Titan, and Enceladus are being planned to take place in the 2020s. All, even Venus, are of astrobiological interest, but honestly none of them is likely to provide that eureka moment of an unambiguous life discovery. That will take time and the analysis of many different types of data. Missions to comets and asteroids are also important to the search for extraterrestrial life because of their chemical and water constituents—even if an individual comet or asteroid is not itself a likely habitat for life.

All these missions are robotic, of course. Humans could explore Mars faster than could even our best rovers. (In fifteen years, the wonderful Opportunity rover travelled a total distance of about forty-five kilometers. I have walked and run that far in a day.[9]) But human missions are much more costly and risky—for the price of a single human mission to one particular (safe) Mars site, we can conduct ten to one hundred robotic missions to multiple (more interesting) sites. As I write this, the Perseverance rover is just beginning its multiyear expedition on the Mars 2020 mission to collect samples for eventual transport to Earth. It is the biggest Mars mission ever, and its results won't be available until the mid-2030s. Bringing samples from Mars to Earth has long been a goal for NASA and planetary scientists.[10] As good as our instruments are when we send them to Mars, the analyses they are capable of are only miniaturized versions of what we can do in Earth-based laboratories. Furthermore, they are, by necessity, about a decade behind current technology, since their planning, testing selection, construction, delivery to the spacecraft, and subsequent delivery to the planet is about a ten-year process. If we can bring chunks of the Martian surface back to Earth and apply the full range of scientific study from many laboratories, many scientists, and many institutions around the world, we will learn much more. If there is life (or evidence of past life) in a Martian sample, it is not very likely we will see it on Mars with our robotic instruments, but it is more probable we will see it if we can bring it back to Earth and slice, dice, and test those chunks with lots of chemical processes and powerful microscopes. But bringing back samples from another planet is more than double the work of taking instruments there—it requires building a rocket to take to Mars and then launching back to Earth. Then, it must be either captured in Earth orbit for eventual retrieval or directly parachuted and landed somewhere where it can be retrieved. The mission involves several spacecraft and is expensive. For this reason, almost all proposals for a Mars sample return mission are international—as is the one being planned by NASA and the European Space Agency, who will aid in the retrieval of the Mars 2020 samples. It is hoped (but not yet fully approved) that the retrieval mission will launch by 2028 and that the samples will get to Earth in the early 2030s.

China, too, is planning a Mars sample return mission. In late 2020, they conducted a lunar sample return—a remarkable achievement on their first try. (The only other nation that conducted a robotic planetary surface sample return was the Soviet Union in the early 1970s, at the height of the space race.) For the Chinese it was a major achievement, one that followed their successful rover mission on the far side of the Moon. In 2021, they conducted their first Mars rover mission, Tianwen-1. They seem committed to planetary exploration and have now developed their own human-crewed space station. Are they getting ready for human missions to the Moon and eventually Mars? Likely so. In chapter 7, I will mention a potential Chinese mission that will exit the solar system. In addition to the U.S. and China, the United Arab Emirates conducted a Mars mission in 2020. Japan, India, and the European Space Agency have also conducted Mars missions. Given the complexity, cost, and fragility of a Mars sample return mission design, it would make sense for it to be conducted with multiple nations participating and contributing, especially if the U.S. and China could cooperate in space.

As noted, there are other solar system targets of astrobiological interest. The most notable is Europa. NASA is planning a mission to be launched in the mid-2020s to orbit Jupiter, with repeated and close flybys of Europa that will enable detailed remote observations of its icy surface—the Europa Clipper mission. The spacecraft will image the surface with cameras and spectrographic instruments and probe beneath it with ice-penetrating radar. It will also seek to study plumes, observed by the Hubble Space Telescope, that may be connected to the ocean beneath the ice-covered surface. It will not have the capability to determine if there is subsurface ocean life there (which is likely to be microscopic, if it exists at all), but it could provide important evidence of potential conditions of habitability there. If only the proposal for the addition of a lander had been supported—it might have peered into or sampled Europa's under-ice ocean. But the cost of adding a lander to an already ambitious orbiter mission was too much for NASA's budget. The European Space Agency is planning an orbiter mission called JUICE—the JUpiter ICy moon Explorer (we just love acronyms). It will repeatedly visit Europa, Ganymede, and Callisto—the three biggest moons of Jupiter.

Another outer planet moon mission is Dragonfly, which will send a rotor-driven aircraft into the atmosphere of Titan that will both sample the nitrogen and organic laden atmosphere and observe the planet's surface. Titan is of enormous interest to astrobiologists, having both water and liquid methane on the surface. Specifically, Titan has an ocean of methane and ethane and lakes of pure methane, though it is far too cold for liquid water—but it has water ice, perhaps brought up there as a liquid by volcanic activity. It is a dynamic world—it is hard to imagine life as we know it there, but it is very much abundant with organics and the constituents of life. Dragonfly is scheduled to launch in the late 2020s. We can hope for a plethora of data about the habitability conditions of Mars, Europa, and Titan to increase our understanding about the probabilities of extraterrestrial life. Scientists also hope to launch a mission to fly through the plumes of water seen erupting from Saturn's moon Enceladus. This will be the easiest water to sample in the solar system (off Earth), since we need no lander—we can just send the spacecraft through the plume and then wipe the water off.[11]

Three Venus missions were very recently approved for launches in the 2020s—two by NASA and one by the European Space Agency. As noted, the astrobiological case for Venus is weak, but a report of a radio-astronomy discovery of a phosphine molecule in its atmosphere sparked speculation about the possibility of life there. Phosphine is a gas released by organics. However, this discovery is highly questionable, and even if the detection is real, it could be a contaminant introduced by ourselves. Still, it is exciting enough to spur these three new missions.

That the search for extraterrestrial life is conducted via the multidisciplinary research of multiple planets (and moons of planets) in our solar system is an important illustration of how science is done. We can't just search for life in our favorite place; we have to piece together observations and experiments on Earth, on other planets and moons of astrobiological interest in our solar system, and on distant worlds in other star systems. We don't know yet whether the universe (or our Galaxy) is teeming with life, but we do know it is teeming with processes relevant to the formation and evolution of life. The discovery of extraterrestrial life may or may not be a eureka moment—it is just as likely to result from a synthesis of studies spanning the globe, the solar system, and the Galaxy. Even if there is a eureka moment, that would only be the beginning. When Copernicus worked out that the planets orbit the Sun rather than the opposite, that was sort of a eureka moment, but it also began a centuries-long inquiry into celestial mechanics and the theory of orbital motion. The mathematics from that theory led to the discovery of more planets and ultimately led to space flight. Similarly, finding life of a kind on one planet will lead to a whole new field of scientific inquiry with new science investigations of other planets and other conditions for life.[12] The vast array of conditions favorable to astrobiology in our solar system makes me think we will still find it here—on Mars, on Europa, someplace. Whether or not we do, we will still want to look farther among those thousands of potentially habitable worlds around other stars, while at the same time looking deeper into the planets of our own solar system.

We'll learn a lot as we search for life in our solar system, and of course we will continue with the efforts here on Earth to synthesize these ingredients into life. We also study extremophiles on Earth—that is, weird forms of life found in niches of extreme conditions, such as embedded in rocks in deserts and deep in the ocean where there is no sunlight and great pressure from the weight of the water. The study of extreme environments on Earth (frigid regions, deserts, undersea vents, etc.) is not just to learn Earth's history, but also to learn about possibilities for life on other worlds. All the varied conditions on the planets of our solar system are also indicators of what to look for as we discover so many more worlds in other star systems of our Galaxy. They are much harder to see and, as I will show in a later chapter, not likely to be explored in the usual sense of us going there—but there may be even more likely candidates for extraterrestrial life around other stars than there are in our solar system (except, of course, for us on Earth).

EXPLORING EXOPLANETS FROM AFAR

It is only in the last twenty-five years that we have been able to actually see planets around other stars—in contrast to the four hundred years we have used telescopes to observe planets in our solar system and stars in the Galaxy, and then the millennia before that in which we observed them with our naked eyes. But, as noted in the previous chapter, we can't see the exoplanets with more than one-pixel resolution—they are basically dots in the night sky, dots bathed in the bright light of their parent stars. Our situation is somewhat similar to that of the ancients who looked at specks of light in the sky and wondered who or what is out there. Of course, we now make spectroscopic measurements and have much higher accuracy instruments for measuring position and velocity, but still the distance to all exoplanets is so great that we can never see more than a dot in a fraction of a pixel.

In chapter 7, we will learn how to overcome this limitation, but even with it we are learning a lot about the abundance of potentially habitable worlds. In this chapter, I will survey the current state of exoplanet observations. We discover and observe exoplanets in three major ways: by the small effect of the planet's mass on its parent star as it orbits the star,[13] by the effect on the star's brightness as the planet moves in front of it and eclipses a small dot, and by direct imaging of the one pixel when the star's brightness can be blocked out. In each of these methods there are actually many different measurements that are sometimes employed, depending on the geometry of the orbits and the masses of the stars and planets. Also important is the star's intrinsic brightness, which is correlated with the star's age. Direct imaging of a planet is still rather rare—it is very hard to do, especially when the planet is relatively small in the presence of a bright star. When exoplanets were first discovered (at the end of the twentieth century and in the first decade of the twenty-first), most were observed due to the effect on the star's radial velocity as the planet moved around it—the planet's gravity affects the star's motion around the center of mass of its system, which in turn shifts the spectrum of the star's light due to the Doppler effect (the effect of relative speed on the wavelength of radiation). In recent years, transit effects have been used and they now dominate the discoveries (accounting for more than three quarters of exoplanet discoveries), as we have been able to more accurately measure the effects on the star's brightness and its variations. Another technique that has accounted for a few discoveries is gravitational microlensing. In a later chapter I will introduce the solar gravity lens as the only way possible to image an exoplanet at high, multipixel resolution. That is something we can only do with a deep-space flight. But microlensing is possible to detect with Earth-based telescopes when fortuitous alignments of stars occur that we can see for a brief time when Earth is in that alignment. If the lensing star has a planet, that effect can be detected, although it is a relatively rare occurrence. Direct imaging of exoplanets has also been possible, in a very few cases with the largest of telescopes, but even in these cases the image is less than one pixel. It would take a telescope with a 3–5 km aperture to fill one pixel with an Earth-sized planet. Direct imaging is very limited—only about one hundred planets of the four thousand so far discovered have been so imaged, and they are all special cases, almost certainly not habitable, because they orbit around very young and hence very bright stars, or because they are very large and in a very distant orbit around the parent star.

The large telescopes on Earth, especially the European Southern Observatory's Very Large Telescope, have contributed significantly to exoplanet discoveries and especially to the direct imaging of them. But we must go to space to really find and observe a lot of them. More than half of those known to date have been observed with the Kepler space telescope, responsible for the rapid increase in discoveries from 2013 to 2018. The Spitzer space telescope, observing at infrared wavelengths, was significant also. It has now been retired, but it is to be succeeded by the Wide-Field Infrared Survey Telescope (WFIRST), now named the Nancy Grace Roman Telescope after the astronomer. The Hubble Space Telescope has made some observations, but the recently operational James Webb Space Telescope (JWST) is expected soon to be a far more significant observer of exoplanets (figure 4).

Successors to Kepler are the Transiting Exoplanet Survey Satellite (TESS; operational since 2018) and a European Space Agency mission, Planetary Transits and Oscillations of Stars (PLATO; under development). We can expect many thousands more discoveries and improved observations, including spectroscopic measurements of Earthlike planets, telling us about atmospheric constituents.

In chapter 4 I noted, "All of the twenty best candidates for potentially habitable exoplanets discovered thus far are more than one hundred light-years from Earth. Within fifty light-years, there are twenty-three terrestrial-sized exoplanets in their stars' habitable zones." That is the current (mid-2021) situation, but as I just noted we can expect thousands more discoveries in the next decade alone, and thus any conclusion about "best" is very premature. The current list of potentially habitable planets is compiled in the "Habitable Exoplanets Catalog."[14] Obviously, this is a dynamic list—it will change with new discoveries and increasing knowledge. There is also a sort of arbitrariness involved in getting on this list—the main criterion is

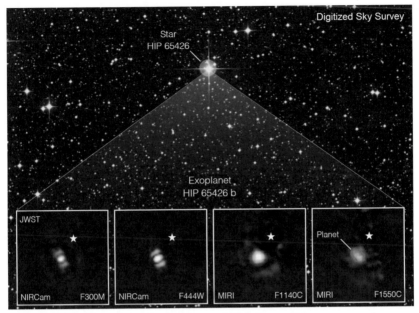

FIGURE 4 These images from the new James Webb Space Telescope are of an exoplanet (HP 65426 b) about six to twelve times the size of Jupiter, taken at four different infrared wavelengths making them appear in different colors. The star shows the location of the parent star, which is blocked out of the image by an internal coronagraph on the Webb telescope.

to be "Earthlike" in size and orbit, or equivalently mass and temperature. It is biased by notions of the habitable zone.[15] The list (table 2), current as of December 2021, is organized according to distance from Earth (Earth is included for comparison purposes).

But note the following:

- The first forty have periods under one hundred days—they are whirring around their star very fast. There is an observational bias that causes our list to be dominated by shorter-period planets—they are the ones whose effects we can notice from multiple observations of orbits. There is no reason to expect habitability favors fast planets going around their stars—Earth isn't fast, and our solar system's fastest planet, Mercury, is surely lifeless. (And probably would be, even if the Sun were cooler, due to stellar dynamics.)

- Only two of the planets on this list orbit a G-type star like our Sun. We know nothing about how planets behave in the environments of other types of stars. Are they dynamically stable enough for habitability?
- Only one of these planets is less than ten light-years away from Earth—all the rest are subject to a lot of observational uncertainty. The one close planet is in a triple star system (Alpha, Beta, and Proxima Centauri), and one must worry about the dynamic stability there in the context of habitable conditions.

It is easy to get excited about all the candidate Earthlike worlds that might be habitable. Twenty years ago, we didn't know of any. Thirty years ago, we debated the very existence and prevalence of exoplanets. But it is also easy to get too excited about these. We are in the barnstorming days of open cockpit aviation, or the Model T days of cars, or the IBM 704 days of supercomputers in terms of where we are at in discovering and categorizing exoplanets. We don't know enough to commit to a billion dollars for any one of them, and it will be a while before we do, given the limits of distance and telescope angular resolution.

TABLE 2 Habitable Exoplanet Catalog List

OBJECT	STAR TYPE	MASS ($M_\oplus$)	RADIUS ($R_\oplus$)	DENSITY (G/CM³)	FLUX ($F_\oplus$)	T_{EQ} (K)	PERIOD (DAYS)	DISTANCE (LY)
Earth	G2V	1	1	5.514	1	255	365.25	0
Proxima Centauri b	M5V	≥1.27	—	4.016	0.7	228	11.186	4.25
Ross 128 b	M4V	≥1.40	—	2.424	1.48	280	9.87	11.03
Gliese 1061 c	M5V	≥1.74	—	—	1.45	275	6.7	12
Gliese 1061 d	M5V	≥1.64	—	—	0.69	218	13	12
Tau Ceti f	G8V	≥3.93	—	3.655	0.32	190	636.1	12
Gliese 273 b	M3V	≥2.89	—	6.45	1.06	258	18.65	12.36
Teegarden's Star b	M7V	≥1.05	—	—	1.15	264	4.91	12.58
Teegarden's Star c	M7V	≥1.11	—	—	0.37	199	11.4	12.58
Kapteyn b	M1VI	≥4.8	—	6.44	0.43	205	48.6	13

TABLE 2 *continued*

OBJECT	STAR TYPE	MASS ($M_\oplus$)	RADIUS ($R_\oplus$)	DENSITY (G/CM³)	FLUX ($F_\oplus$)	T_{EQ} (K)	PERIOD (DAYS)	DISTANCE (LY)
Wolf 1061 c	M3V	≥3.41	—	5.79	1.3	271	17.9	13.8
Gliese 832 c	M2V	≥5.40	—	—	0.99	253	35.7	16
Gliese 229 A c	M1V	≥7.27	—	—	0.53	216	122	18.8
Gliese 625 b	M2V	2.82±0.51	—	—	—	—	14.628	21.3
Gliese 667 C c	M1V	≥3.81	—	5.603	0.88	247	28.1	23.62
Gliese 357 d	M2V	≥6.10	—	2.617	0.38	200	55.7	31
Luyten 98–59 f	M3V	2.46	—	—	>1	~280	23.15	34.648
Gliese 180 c	M2V	≥6.40	—	—	0.78	239	24.3	39
TRAPPIST-1d	M8V	0.3	0.78	3.39	1.04	258	4.05	39
TRAPPIST-1e	M8V	0.77	0.91	5.65	0.67	230	6.1	39
TRAPPIST-1f	M8V	0.93	1.05	3.3±0.9	0.38	200	9.2	39
TRAPPIST-1g	M8V	1.15	1.15	4.186	0.26	182	12.4	39
HD 40307 g	K2V	≥7.09	—	2.855603	0.67	226	197.8	42
Gliese 163 c	M3V	≥6.80	—	—	1.41	277	25.6	49
LHS 1140 b	M4V	6.98	1.73	$7.82^{+0.98}_{-0.88}$	0.5	214	24.7	49
Gliese 3293 d	M2V	≥7.60	—	—	0.59	223	48.1	66
TOI 700 d	M2V	~1.72	1.14	5.631	0.87	246	37.4	102
K2–288 B b	M3V	—	1.91	—	0.44	207	31.4	214
K2–72 e	M?V	~2.21	1.29	5.675	1.11	261	24.2	217
K2–9 b	M2V	—	2.25	—	1.45	279	18.4	270
Kepler-1649 c	M5V	—	1.06	5.54	0.75	237	19.5	301
K2–296 b	M?V	—	1.87	—	1.15	264	28.2	519
Kepler-186 f	M1V	—	1.17	—	0.29	188	129.9	579
Kepler-22 b	G5V	—	2.38	—	1.1	261	289.9	635
Kepler-737 b	M	4.5	1.96	—	—	—	28.5992	669
Kepler-296 e	K7V	—	1.52	—	1.5	276	34.1	737

TABLE 2 *continued*

OBJECT	STAR TYPE	MASS ($M_\oplus$)	RADIUS ($R_\oplus$)	DENSITY (G/CM³)	FLUX ($F_\oplus$)	T_{EQ} (K)	PERIOD (DAYS)	DISTANCE (LY)
Kepler-296 f	K7V	—	1.8	—	0.66	225	63.3	737
Kepler-1540 b	K?V	—	2.49	—	0.92	250	125.4	799
Kepler-1652 b	M?V	—	1.6	—	0.84	244	38.1	822
Kepler-1229 b	M?V	—	1.4	5.426	0.49	213	86.8	866
Kepler-705 b	M?V	—	2.11	2.994	0.83	243	56.1	903
Kepler-62 e	K2V	—	1.61	—	1.15	264	122.4	981
Kepler-62 f	K2V	—	1.41	5.509	0.41	204	267.3	981
Kepler-440 b	K6V	—	1.91	—	1.44	273	101.1	981
Kepler-61 b	K7V	—	2.15	3.6	1.39	273	59.9	1092
Kepler-1544 b	K2V	—	1.78	—	0.9	248	168.8	1092
Kepler-26 e	K	—	2.1	—	—	—	46.8	1104
Kepler-442 b	K?V	—	1.35	5.272	0.7	233	112.3	1194
Kepler-1410 b	K?V	—	1.78	—	1.34	274	60.9	1196
Kepler-174 d	K3V	—	2.19	—	0.43	206	247.4	1254
Kepler-283 c	K5V	—	1.82	—	0.9	248	92.7	1526
Kepler-298 d	K5V	—	2.5	—	1.29	271	77.5	1689
Kepler-452 b	G2V	5	1.63	—	1.11	261	384.8	1799
Kepler-1701 b	K?V	—	2.22	—	1.37	275	169.1	1904
Kepler-1632 b	F?V	—	2.47	—	1.27	270	448.3	2337
Kepler-1653 b	K?V	—	2.17	—	1.04	258	140.3	2461
Kepler-1552 b	K?V	—	2.47	—	1.1	261	184.8	2507
Kepler-443 b	K3V	—	2.35	2.888859	0.89	247	177.7	2615
Kepler-1606 b	G?V	—	2.07	—	1.41	277	196.4	2710
Kepler-1090 b	K0V	—	2.25	—	1.2	267	198.7	2800
Kepler-1638 b	G4V	—	1.87	—	1.39	276	259.3	4973

Credit: NASA

The Habitable Exoplanet Catalog defines a factor called the Earth Similarity Index that compares the stellar flux, mass, and/or radius of the observed exoplanet to that of Earth—that is, it is a (theoretically deduced) measure of the capability of sustaining water and an atmosphere. The top five according to this factor are Teegarden's Star b, TOI 700 b, K2–72 e, Trappist-1 d and Kepler-1649 c (figure 5).[16] There is no point in picking a favorite, or basing any priority on this information (although people will)—the substantive point is that, even at this early stage in the first generation of space observations, we have many interesting potentially habitable exoplanets to explore—and we will have many times more.

There are proposals for another generation of space telescopes after these, and it is reasonable to expect that within the next decade some very

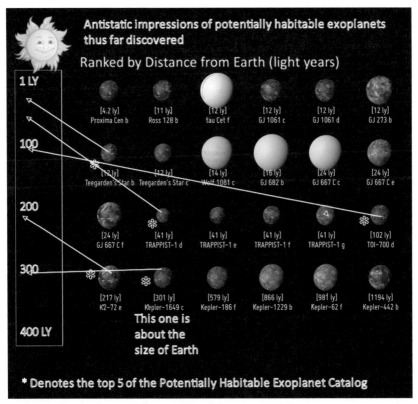

FIGURE 5 The top five potentially habitable planets selected from the Habitable Exoplanets Catalog. Courtesy of the Planetary Habitability Laboratory, Univ. Puerto Rico at Arecibo, with annotations by the author.

promising candidate habitable planets will be identified—perhaps even some suspected of being inhabited with plant or primitive cell life (based on coloring and spectroscopic measurements). The situation will be somewhat analogous to the discovery of water on Mars—it will likely occur after years of accumulating inferences from better and better data, but without definitive evidence until directly observed. We will only get hints. No observation from Earth, or even from space in our solar system, will resolve an exoplanet even to one pixel. No features can be seen. There are only two ways to do that—go there (interstellar flight) or use the one hundred billion magnifying power of the solar gravity lens. These are addressed in the next two chapters.

6

VISITING EXOPLANETS: INTERSTELLAR FLIGHT, A BRIDGE TOO FAR

Space is too big and our lives too short to reach the stars—
at least physically.

THE LURE OF interstellar travel is large. But larger yet is the scale of interstellar space. The nearest star system, Proxima Centauri, is about 4.5 light-years from the Sun. That equals about 270,000 astronomical units (AU), or forty trillion kilometers (twenty-five trillion miles).[1] That is the nearest star. If your goal is to investigate a promising, potentially habitable exoplanet, one that has been identified to have the potential for life, then you should set your sights to reach one of the ten nearest stars with identified potentially habitable planets. That takes you at least four times farther to about seventeen light-years, or about one million AU (see table 2 in the previous chapter). When enthusiasts talk about interstellar travel, they usually focus on the nearest star rather than the interesting ones—the ones that are more likely to have life on one of their planets. The furthest spacecraft we have ever sent has now reached about 150 AU (in forty-three years), about 0.0005 of the way to Proxima Centauri and 0.000140 of the way to seventeen light-years. That is Voyager 1—the farthest we have gone, and the fastest, at about 3.2 AU per year.

That said, interstellar flight is easy if your species is a patient one. Voyager will reach the distance of the nearest star in about one hundred thousand years[2]—roughly the same age as the human species. But interstellar flight on a human time scale is nearly impossible. To reach the nearest star in (for example) one hundred years would require an average speed of about 2,700 AU per year (equivalent to a flight time to Mars of about seven days). That is 12,800 kilometers per second (28.6 million miles per hour). The largest rocket with chemical propulsion can produce a speed of about five kilometers per second,[3] the largest electric propulsion plasma rocket can go three thousand kilometers per second. If you used the whole mass of Earth as fuel, the electric propulsion could send one quintillionth of a gram on the voyage—and the chemical propulsion could send nothing. What about nuclear? The definitive serious study to try to figure that out, called Project Orion, was done in the 1960s. One of its leaders was the earlier mentioned brilliant physicist and big thinker Freeman Dyson of the Institute for Advanced Study in Princeton. Orion assumed the use of nuclear fusion rockets and envisioned using three hundred thousand one-megaton nuclear bombs continuously exploding for rocket propulsion. The spacecraft would reach Alpha Centauri in 133 years, flying at about 3 percent the speed of light. That is not very practical. The British Interplanetary Society, a long-time champion of interstellar flight studies, sponsored another nuclear propulsion study called Project Daedalus to send a scientific payload on a two-stage fusion rocket to Barnard's Star (about six light-years away).[4] The total mass of the vehicle was a half-million tons, and required mining helium-3 for fuel from the atmosphere in order to get going. It is a brilliant theoretical study, but still not practical.

Even more theoretical (and less practical) is to use pure energy from a matter-antimatter engine (as in *Star Trek*). Theoretically, it creates the largest possible energy for any application. The energy produced by combining matter and antimatter could provide a velocity of thirty-five thousand kilometers per second—which would get us to the nearest star in four hundred years. That might be acceptable compared to a human lifetime (especially if we keep increasing the human lifetime) except for one problem—there isn't enough antimatter in the world to make such an engine and making antimatter ourselves would take thousands of years—as far as we know now. So far, we have made about one gram. There is another problem: even if we had the antimatter, we have no idea how to contain it and feed it into an engine. It annihilates everything it touches.

There is simply no way to propel a spacecraft that has to carry its own fuel to another star system in hundreds (or even thousands) of years. The more fuel you need—even if it is pure energy fuel like antimatter, the heavier are the rockets and the more impossible it is to get anywhere fast enough. This is expressed mathematically in the rocket equation, first derived by the father of space flight, Russian scientist Konstantin Tsiolkovsky. It describes how fast a rocket can go as the product of the rocket's exhaust velocity (determined by the type of fuel) and the logarithm of the ratio of its initial mass (including fuel) and final mass (payload and structure). That means that the mass of the fuel required goes up exponentially the faster we try to go. Engineers refer to the problem as the "tyranny of the rocket equation." We can't build a rocket (with any fuel) to go fast enough for interstellar travel within a human lifetime, or two, or ten.

We must get our propulsion outside the rocket. The only way possible might be to take no fuel and instead get our power externally. The Sun is a source of power, and using it on a solar sail we can indeed build up enough speed to exit our solar system.[5] We do this by going close to the Sun with a large sail and using the momentum transfer by the photon energy of the Sun to provide thrust to the vehicle. The amount of thrust is proportional to the area of the sail divided by the mass multiplied by the solar power. But the practical limit on solar sailing is around 50 AU per year, even with huge sails and very close solar distances—a far cry from the necessary 3,000 AU per year. We can replace the Sun with light power from a large laser array focusing its beam on the sailcraft. We can make the laser as powerful as we want, and laser light can be narrowly focused to put all that power onto a sail. Laser sails seem to be the only (barely) practical means of fast interstellar travel utilizing known technology. The concept was first invented by Robert Forward, a brilliant physicist who provided engineering and science analysis in professional literature and also wrote about it in science fiction literature. His work has now been updated and expanded in academic studies by Phil Lubin and is the basis of a notable privately funded study called Starshot to create an interstellar flight in the twenty-first century, funded by a billionaire science enthusiast, Yuri Milner.

Lubin's work shows that to send one kilogram to the nearest star in one hundred years requires about a two-hundred-gigawatt laser array. The size of that array would be greater than 250 square kilometers. That is sixty thousand times the highest-power laser array now in existence and nearly ten times the output of the world's largest power-generating station, the Three Gorges Dam in China. If we wanted to send a one-hundred-kilogram payload with that same power, it would take over four hundred years. The

Starshot study has chosen the other direction—to send only one gram (!) to accomplish the flight in twenty years.[6] There are many other ideas for interstellar travel, mostly in science fiction literature (some of these are discussed below), but this is the only one now underway with credible scientific and engineering participation that is potentially considered practical. But "credible" and "practical" (like beauty) are in the eye of the beholder. I personally question whether sending one gram pushed by two hundred gigawatts of lasers on a twenty-year mission to fly by the target exoplanet at sixty thousand kilometers per second to obtain four seconds of data from within a lunar distance of the planet is either credible or practical. Especially when one thinks of their optimistic budget of $10 billion. The four seconds of flyby data is not just a limitation of Starshot's design, it is intrinsic to any notion of a fast interstellar mission to an exoplanet. Passing by an exoplanet at this fraction of relativistic speed doesn't give much observing time. Remember how wrong a picture we got of Mars from the Mariner 4 first flyby of the planet. We thought it was dead and barren like the Moon. It wasn't until we got orbiters there and could make repeated and synoptic observations that we learned what the planet really was like. Flying by an exoplanet at relativistic, interstellar flight speed—even if we can do it—is likely to lead to a similar wrong picture. It is certainly not going to tell us anything about life on that world. That's a huge problem, especially when we consider that interstellar flight would be only to the nearest star. The Starshot study is totally focused on the Alpha Centauri system; in fact, buoyed by the discovery of an Earth-sized planet around Proxima Centauri, they designed their mission to go to that planet. But Proxima Centauri is part of a triple-star system likely subject to wild dynamics that would affect any potential climate stability, which is not promising for life as we know it. Furthermore, recently scientists have observed a huge solar flare there that would have likely wiped out any life on such a planet had it even been there.[7] Thus, my negativity on the interstellar flight goal is less about how difficult it is technically and more about how little we would learn and how superficially fast we would fly by any target.[8] In later chapters I will suggest that there is a remote-sensing alternative for closeup observation of exoplanets, offering advantages analogous to those of orbiting a planet instead of merely flying by it.

However, we should still be open-minded and careful not to arbitrarily rule out the possibility of new scientific discoveries or new technologies that might change our thinking. As hard as it is to get away from the numbers and the scale of space, there are a few topics in theoretical physics that have been used in science fiction that I should at least mention. Sometimes science fiction is a harbinger of things to come; more times it is not. One idea is warp drive—a theory that claims it is possible to contract space-time (Einstein's expression for the gravity of the universe) in front of a spacecraft and expand it behind to create a means for faster-than-light propulsion. Its use was proposed by physicist Miguel Alcubierre. It is neither proved nor disproved, but no one has been able to figure out a means to provide the power necessary to produce such an effect. It probably breaks the laws of physics, although some claim it gets around them by altering space-time, perhaps with something such as negative energy—another concept that has no definitive proof. Harold (Sonny) White, formerly of NASA, has proposed that a warp drive (or negative energy) can be created by sending powerful oscillating electromagnetic fields through a superconducting torus. NASA has attempted a very small-scale laboratory experiment to test the creation of a warp drive, without definitive results to date.[9]

Another theoretical idea is to fly through wormholes. The theory of general relativity has a mathematical solution called the Einstein–Rosen bridge that connects two different spots in space-time instantaneously. In theory, that would permit not just faster-than-light travel, but also time travel. (Which is why it is also probably impossible.) The physicist Kip Thorne used this idea to help Christopher Nolan create the movie *Interstellar*.[10] Wormholes are also sometimes described as a fourth dimension,[11] whereby we in the three-dimensional world can shortcut to another location in a way analogous to two-dimensional people being able to move on their plane by jumping into a third dimension. That analogy is strained by the fact that the four-dimensional reality we live in has three spatial dimensions, while the fourth dimension is time. It seems possible that wormholes exist—cosmological and gravity theories allow it. But it seems far less likely that they can actually be used for space (or time) travel since, even if they do exist, they are very unstable and depend on negative energy (again) and exotic matter, now only a subject of speculation.

Of course, interstellar travel is possible, even easy, if we are willing to take a lot longer than a human lifetime. "Generational starships" are frequently used in science fiction. This has no theoretical problems, but it is neither very practical nor very satisfying. A robotic spacecraft could do this with nuclear power—as the Voyager spacecraft are now doing. However, their nuclear power will run out long before they encounter another star system. Future robotic spacecraft with AI and self-replicating robots might be able to fly to interstellar destinations, but if the point is to communicate their findings to the humans who sent them, that is not going to happen. A different point of view might be that they are the successors to human evolution (as discussed earlier in this chapter), and some think that is possible. For humans to fly on a spacecraft for multiple generations, it would require the spacecraft to be a totally self-sufficient ecosystem and probably a lot of AI in the starship computers. The ships would be huge, big enough for a self-sufficient biosphere and with a lot of nuclear power and raw materials. They might also derive energy and fuel from the various star systems they putatively pass through. To imagine these, we must resort to science fiction and very long time scales. The brilliant science fiction author Kim Stanley Robinson used the idea of generational starships in his novel *Aurora* (Orbit, 2015) to describe a twenty-fifth-century effort to spread humanity into the Galaxy. Several thousand people were sent on a voyage to colonize an identified habitable planet in the Tau Ceti star system.[12] The book's premise was that it was technically feasible, but still the effort failed because humans could not adapt to non-Earth conditions, no matter how advanced the engineering. Our DNA prevented it. The interstellar voyagers, in another multigenerational space flight, had to return home—back to Earth, the only place where humans could thrive. Robinson's view that interstellar flight is a pipe dream meshes well with that of the biologists Ernst Mayr and E. O. Wilson, both of whom I cited earlier.[13] They exemplify the point made in an earlier chapter about the differing world views of astronomers and biologists about extraterrestrial life. This notion that our biology is so unique among the stars that it inhibits interstellar flight for humans and extraterrestrial intelligence in the universe is, if it is true, profound and worth dwelling on. As I noted, intelligence, as a product of evolution on Earth, is very recent and perhaps very short-lived. Not only do we have to consider all the existential threats to civilization—pandemics, asteroid impacts, nuclear war, climate change, extreme biodiversity reduction, misprogrammed AI, or misprogrammed genetic engineering—we should also speculate about just how humans will evolve. Even without the existential threat of misprogramming, the development of AI technology and the integration of biology and electronics could lead to a gradual shift from carbon-based life to silicon-based something. (Whether to call the silicon-based, noncarbon AI robot "life" or not is something to debate—see chapter 3). Robotic flight and AI are probably not inhibited from travelling to the stars on long time scales that outlast the human species. This brings up both the notion of our future evolution into AI and the idea of von Neumann probes.[14] These are self-replicating probes that can fly indefinitely—an idea that may have seemed preposterous when John von Neumann described it in the 1920s and 1930s, but which is very believable now, with 3D printers being sent on spacecraft and AI programming advancing. Von Neumann probes certainly could be sent on interstellar voyages, perhaps even by intelligent species who die out before the probes get anywhere. There is a lot of scope for speculation here, but then we still have to wrestle with the question posed in chapter 3—Where are they?

To be fair, a contrasting science fiction idea about the impossibility of human adaptation to interstellar worlds is offered by Andy Weir in his novel *Project Hail Mary* (Ballantine Books, 2021). In it he describes a desperate human voyage to another star in an attempt to save Earth from a plague of organisms being manufactured on the Sun and rapidly decreasing the Sun's energy. At the destination star the human traveller meets another interstellar voyager from a different planet on a similar mission. But the nonhuman life-form is silicon based with extraordinary manufacturing capabilities that complement the human brain. The human and nonhuman work so well together that ultimately (after they save their worlds) the human goes to live on the nonhuman world and adapts even though it has phenomenally un-Earthlike conditions. It is a clever story, but it does not add at all to the belief that interstellar travel is possible for humans. Many science fiction writers have come up with brilliant scenarios for interstellar voyages, but they always seem to rest on at least one impossibility.

Finally, we come to an idea for interstellar travel that I find especially intriguing—hopping on an interstellar asteroid. In 2013, Joseph Breeden published a paper that described how a binary asteroid (with two parts co-orbiting) might be redirected to a sun-grazing orbit and then one part dropped off to orbit the Sun, causing the other to benefit from both the resulting angular momentum transfer and the gravity assist at close approach to the Sun and fly off at interstellar speed.[15] Breeden notes that this dynamical behavior has been observed in the motions of binary stars.

The two ideas Breeden made use of were the redirection of an asteroid by artificial means and the large change in velocity (a high Delta-v) imparted to a body when making a maneuver at low perihelion (closest approach to the Sun). The first, developed as part of a novel concept for NASA's human spaceflight program, is for astronauts to use a redirected asteroid placed near the Moon as an interim target for a voyage to Mars. The asteroid target would be more practical than the conventional idea of landing on the Moon, since no lander system would have to be built. The Obama administration supported this approach, but the Trump administration cancelled it and added the lunar landing stop back into the NASA human spaceflight plan. The asteroid redirection would have been accomplished by a robotic electric propulsion spacecraft.

The other idea, known in celestial mechanics as the Oberth maneuver, is based in orbit dynamics.[16] The most efficient way to expand an orbital ellipse is with a Delta-v at perihelion—with enough Delta-v, the ellipse expands into a hyperbola with a nearly straight-line trajectory exiting the solar system.[17] Breeden states that the part of the binary asteroid that flies off at interstellar speed could contain a crew or maybe some DNA. I really liked the suggestion because, as the reader can tell from my earlier comments in this chapter, I don't see any other practical way for interstellar travel. Calling this idea practical is a bit of a stretch considering how close the object must get to the Sun and how extremely sensitive are the orbit dynamics. Harnessing this for human spaceflight seems a stretch. Breeden also does not deal with the engineering details of creating the binary asteroid or turning part of it into a spaceship, or how to "drop off" the part to go into solar orbit. It is clearly not practical for us now, and it too would involve very long flight times—thousands or even millions of years. But an advanced civilization might be capable of using this method. I also wonder whether such a process could occur naturally—that is, with a natural binary asteroid (there are such things) coming close to the Sun and getting ripped apart as it grazes the Sun. In 2017, an interstellar object (ISO) was observed for the first time as it passed through the solar system. It was subsequently named 'Oumuamua. It is weird looking—at first thought to be shaped like a cigar (long and thin) and more lately described as cookie shaped, or maybe flying saucer shaped (figure 6). That might be laughable, except that in 2021 Prof. Avi Loeb, a distinguished astronomer at Harvard University and former chair of the National Academies Board on Physics

FIGURE 6 Artist's concept of 'Oumuamua, discovered in 2017, an ISO with a strange shape and orbit. Courtesy of the European Southern Observatory / M. Kornmesser.

and Astronomy, published a paper and then a book concluding that 'Oumuamua is indeed an alien spaceship.[18] His reasoning is strange (for a scientist)—namely that the object's orbital behavior is full of anomalies that are not readily explained. His explanation is that it is an alien spaceship. He invokes Occam's razor—the principle that in the absence of any contradicting information the simplest explanation is usually the best. Personally, I think he has it backwards—a much simpler explanation is that it is a natural errant asteroid or comet from a peculiar, but still natural, orbit with known dynamical forces (such as a binary asteroid or a comet of unusual composition). An even more natural explanation has been published by two planetary scientists, who conclude that 'Oumuamua is likely a nitrogen ice fragment from a Pluto-like object in another star system that resulted from an impact on that object which ejected the fragment at interstellar velocity.[19] That's extraordinary, and the binary asteroid hypothesis is extraordinary, but such things have been observed and they are far less extraordinary than alien spacecraft. Loeb assumes the ISO is not just a spacecraft but one with solar sails that give it its initial speed—and that somehow still function as the object flies for tens of thousands or millions of years through interstellar space. If it is an alien spacecraft, I think Breeden's redirected and controlled binary asteroid propulsion method is more likely. But to be clear, I don't think it is an alien spacecraft.

All in all, interstellar flight is a bridge too far for humans, and even for their robots on any human time scale. Once again, we are alone—not just in terms of biology, but also distance. If advanced (and still unknown) technology could package our brains into computer systems with communications and power weighing just one kilogram, it would still take one hundred years to reach the nearest star. Getting to the very nearest (and likely unhabitable) exoplanet with extreme technologies not yet realized would still yield little information as the spacecraft whizzes by the exoplanet in a few seconds. We know there are many planets to explore: potentially habitable planets around other stars, along with the key places of astrobiological interest in our solar system. In our own solar system, we will reach them—even land and rove on them—but beyond our solar system, our exploration will be remote. However, observations of a remote object light-years away won't really reveal much, and what it does reveal will be ambiguous. With so many potentially habitable exoplanets to explore, we need to make our exploration virtual—that is, to get enough data to allow us to investigate those worlds at high resolution. We can't get there for real exploration, but Nature (and Einstein) has provided us with a tool for virtual exploration—to explore many worlds in detail. In the next chapter I describe how we can extend our vision to other solar systems.

1

USING NATURE'S TELESCOPE

Nature (and Einstein) has provided us with a scientific instrument capable seeing exoplanets—up close and personal. It is far away, but within our reach.

REMEMBER THE REACTION in 1996 to the putative microscopic evidence of life discovered in a Mars meteorite? The meteorite was discovered in Antarctica where it had landed thousands of years ago, and upon analysis it was conclusively found to have come from Mars. The principal piece of evidence was that it had gases imbedded in the rock that were exactly and uniquely in the combination of those gases in the Martian atmosphere. Years after its discovery and identification as Martian, a microscopic examination of the inside of the meteorite revealed what very much looked to be the remnants of living cells. The discovery caused a sensation—it was announced by President Clinton as a momentous discovery, one that had to be followed up with further exploration. In the aftermath, a series of Mars missions was approved—landers and orbiters to be sent at nearly every Mars launch opportunity. An international commitment was made to send a spacecraft to Mars to collect samples on its surface and return them to Earth. To (almost) everyone's credit, the general view of the discovery was that it must be verified, followed up, understood, etc. The scientific analysis was studied and debated carefully—both with a lot of skepticism and admiration for the work done by the authors of the published results. The evidence for life on Mars is now considered more uncertain, and, in fact, the preponderance of scientific opinion is that whatever is in that meteorite is likely not life. But the drive for continued exploration of Mars continues, as it should. Mars is still likely to have had life in its warmer and wetter past, and it is possible that some microorganisms from that past survive today buried in its interior. As our nearest neighbor with a possibility of life in its past, present, and even in its future if we get there, Mars is a high priority for space exploration. Both the U.S. and China landed there in 2021, the United Arab Emirates conducted an orbiter mission there in 2020–21, and Japan and India are planning orbiters there to be launched in this decade. Beyond that, the U.S. and Europe have begun development of a sample return mission from Mars, as has China. Life on other worlds is an Earthly interest, and exploring other worlds is an Earthly endeavor. As I have noted in earlier chapters it is ever more likely that evidence or at least hints of life on exoplanets will be found in the coming years—but (as with Mars) it will be, at best, only ambiguous evidence, since we can't go there or get more than a single-pixel image of an exoplanet. We are able to send landers with robotic laboratories to Mars repeatedly, and we will be able to bring back samples to analyze in large facilities on Earth. With exoplanets, we can't do that. I detailed in the previous chapter that going to a potentially habitable exoplanet is a bridge too far: either the trip would take too long, or the spacecraft would whiz by the target so fast as to make observations not useful. That is, even if we could solve the propulsion and spacecraft engineering problems to carry a reasonably sized science payload. Furthermore, the most interesting exoplanets are not likely to be the closest ones—making it even more impossible to get to one with putative life. As for building bigger telescopes, it would take a telescope tens of kilometers in diameter, in space, to resolve the nearest exoplanet, even with one pixel. We cannot, for any reasonable cost, build a telescope large enough to resolve features on an exoplanet. The largest single-mirror optical telescope on Earth (expected to have first light in 2023) is the Large Synoptic Survey Telescope (LSST; now the Vera C. Rubin Observatory) in Chile—with an 8.4-meter (28-foot) aperture. The resolving power of a telescope is given by the wavelength of the light divided by the telescope diameter. For the shortest visible wavelength, the wavelength is 400 nanometers. Dividing that by 8.4 meters gives an angular resolution of 47.6 nanoradians (about twenty-millionths of a degree). Multiplying that by the distance to the nearest exoplanet (forty quadrillion meters), you get a resolution of 1.9 million kilometers, about

250 times the size of Earth. That is, the world's largest telescope would not be able to resolve a whole Earth-sized planet. A much larger telescope, the Extremely Large Telescope, is under construction—also in Chile. It will be 39.3 meters—4.6 times larger than the LSST. Its resolution of the nearest exoplanet would be 0.4 million kilometers—still more than fifty times the size of Earth. It would take a telescope a thousand times bigger than the LSST to resolve an Earth-sized planet. And, of course, that telescope on the ground is a lot bigger than anything we could put in space where the seeing is much better. We are many orders of magnitude away from being able to resolve or see any details on an exoplanet.[1]

Fortunately, however, Nature (and Einstein) provides a solution. There is a "natural" telescope in our solar system that provides a magnification of one hundred billion power. All we have to do is get to it.[2]

When Albert Einstein published the theory of general relativity in 1915, he included certain predictions that could be tested by accurate scientific measurements. One prediction was that light rays, which are made up of zero-mass photons, would be bent by gravity. As light rays pass by a massive object, such as the Sun, they would deviate from the Newtonian straight line on which they were thought to travel onto an Einsteinian curved line, bent by the gravitational field in which they were travelling. In 1919, the British astronomer Sir Arthur Eddington measured that bending of starlight by the Sun, exactly as predicted by the theory. This measurement (and many, many others since) confirmed the new theory and the genius of its author. The bending of light rays as they pass close to the Sun means that the Sun acts like a lens, curving the rays from a distant object and focusing them at a certain point that depends on the mass of the Sun and the radial distance of the light ray from the limb of the Sun's disk. Light rays from a distant object (light-years away) will pass by the Sun's limb at varying distances from it, so the focus is not just a point, but a series of points—that is, a focal line that extends indefinitely outward away from the Sun. Plugging in the numbers, one can calculate that the focal line begins 547 AU away from the Sun and extends outward on a straight line formed by the distant object and the Sun. The intensity of light along the focal line is amplified by one hundred billion! This solar gravity lens (SGL) provides an effective one hundred billion magnification, which means that a distant object could be resolved to a scale of a few kilometers. If that distant object is an exoplanet, that would mean we could see regional features on its surface the size of continents, seas, lakes, forests, and ice caps[3]—not just on the nearest exoplanet, but on scores of potentially habitable planets within the radius of hundreds of light-years from Earth. All we have to do to get high-resolution images of potentially habitable exoplanets is to send spacecraft with relatively small telescopes (a meter or two in diameter) out on to the target exoplanet's focal line. The spacecraft does not need to stop at a focal point, but rather follow a path along the focal line. That's all—just send the spacecraft to 547-plus AU to travel straight down the focal line and collect the image of the exoplanet.

There are a few complications:

- The Sun's limb isn't sharp; its corona makes it fuzzy due to dynamic emissions, the result of which is to extend the practical first focal point where an image can be captured to more than 547 AU, to perhaps 625–650 AU.
- The image is a light field projected by the SGL on image planes normal to the focal line. In the focal region the spacecraft telescope looks back toward the exoplanet (and thus toward the Sun) and it sees an annulus known as the Einstein Ring formed around the Sun by the light from the exoplanet. The diameter of the ring is slightly larger than the Sun (millions of kilometers), and the thickness of the annulus is a few kilometers. This requires the spacecraft to sample the image pixel by pixel.
- The spacecraft must not just reach the focal line, it must then fly along it as it extends outward, with some maneuverability to capture the pixels in the Einstein Ring and integrate them into a deconvolved image over a period of weeks, months, or even years.
- The Sun is not perfectly round, which introduces a kind of aberration in the SGL—worse for planets in the solar equatorial plane and not so bad for those at higher solar equatorial latitudes.
- And, of course, the spacecraft has to get to 650-plus AU in a reasonable amount of time—say the working lifetime of the mission scientists, or roughly twenty-five to forty years. That works out to an average speed of 20–30 AU per year. The fastest spacecraft we have ever sent out of the solar system is the Voyager spacecraft, travelling at about 3.2 AU per year. Seven times faster is hard, and it is fast (approximately 265,000 miles per hour)—but it is a lot less hard and fast than the speed of about 13,000 AU per year, which is required for interstellar flight to actually reach (only) the nearest star within twenty years. Compared to interstellar flight, it is six hundred times slower and about a million times easier.

As described above, the imaging is more complicated than snapping pictures with a camera, as we now do routinely for the planets in our solar system. Slava Turyshev of NASA's Jet Propulsion Laboratory, together with his Canadian colleague, Viktor Toth, has spent several years analyzing the wave properties of light passing through the solar gravity lens and coming up with a technique for capturing the images. The image, as seen by the spacecraft's telescope, is projected along the Einstein Ring. Each pixel from the exoplanet is placed in the image plane and creates the brightness of the Einstein Ring. For a 6,000 km radius planet (the approximate size of Earth), the width of the annulus is about 1.3 km—at the beginning of the focal line. As we move outward, the width of the annulus becomes larger. By the time we get to 900 AU, it is about twice larger. Figure 7 shows the light rays of an Earthlike planet (A) being bent as they pass the Sun to form the Einstein Ring (B) around the focal line emanating outward from 547 AU. The resulting image is a convolution of the true source image with that formed by the SGL (C) and must be deconvoluted by image processing—data collection that takes place over many months as it is accumulated from the rotating planet. At the beginning, the brightness data collected looks like the picture on the left (C), but after sufficient integration time and deconvolution it looks like that on the right (D)—showing the features of the Earthlike planet.[4] It still takes a telescope to get this image—but one that fits on a small spacecraft—one to two meters in diameter. In addition, a coronagraph is needed to block out the Sun, which of course is in the center of the picture. A coronagraph blocks out the central light internally within the telescope by computing the solar disk's location. It won't be perfect, because the Sun has a dynamically varying corona, but with sufficient integration time that can be modelled as well. To collect the data, the spacecraft with its telescope must be maneuvered over the distance of thousands of kilometers to collect all the data necessary to construct the image, pixel by pixel. In addition, everything in space is moving: the Sun around the solar system's barycenter, the exoplanet around its host star, and the star relative to the Sun. All these motions must be tracked and require the spacecraft to have microthrusters and electric propulsion to move the thousands of kilometers in the focal region. Way out at 600-plus AU from the Sun, there will be no solar power—so we will need onboard nuclear power for both the electric propulsion and operation of the spacecraft instruments and computer (such as that on the Voyager spacecraft), say with a radioisotope thermoelectric generator.[5] This is quite manageable for the small maneuvers needed to capture the exoplanet's image. The bigger propulsion job is just to get to that focal region—to get our spacecraft to fly out of the solar system at greater than 20 AU per year.

As stated earlier, the focal region, the SGL focus, is a line beginning 547 AU from the Sun and extending indefinitely outward. The line's other point is the exoplanet (or star), which is the imaging target, on the other side of the Sun. A distance of 547 AU is well into the interstellar medium, which is defined as the region whose particles are primarily those streaming from distant stars. It begins (irregularly) at around 120 AU from the Sun with a shock wave (known as the heliopause), which is the result of the collision of the solar wind and interstellar particles. The interstellar medium is depicted in figure 8.

Note that the scale in figure 8 is logarithmic—otherwise the distances in the solar system wouldn't show up on a map depicting distances in light-years. On this scale, the SGL focus appears roughly halfway to the nearest star; actually, it is about 1/500 of the way (0.2 percent). Even more noticeable is that there is not much else on the way. The interstellar medium is almost empty—with about an average density of one atom per cubic centimeter. The interstellar medium itself is an important area for scientific study. Heliophysicists want to understand the composition and the dynamics of the wind that blows here from other stars in our Galaxy. Only the Voyager spacecraft have made it into the interstellar medium from Earth. Scientists would like to get two to three times farther to sample that stellar wind separate from the solar wind of our own solar system. There is something missing on figure 8, however: Planet X. That's because we don't know where it is, or even if it is. But if it is there, then there will be another destination for planetary scientists to explore in the interstellar medium. There has long been speculation about a planet X (for unknown) so far distant that we can't see it. In 2015, scientists at Caltech noticed unusual behavior among Kuiper belt object orbits that tended to clump together. They found that hypothesizing a Neptune-sized planet in a very elliptical orbit very far out (>600 AU) helped to explain the KBO orbits.[6] The gravitational effect of such a Planet X could affect the KBO objects in a way similar to the observed behavior. This is not proof that it is the cause, or that there is a Planet X—it's just suggestive. The scientists referred to it as Planet 9 (annoying Pluto diehards who still want their planet to be #9). Astronomers are searching for this planet—but there is way too

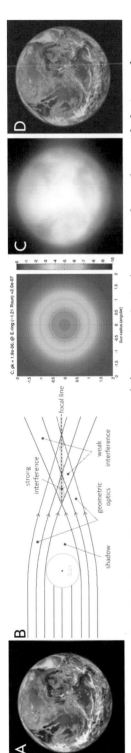

FIGURE 7 The solar gravity lens: light from a distant Earthlike exoplanet (A) is bent by the Sun's gravity to form a lens, with the image data captured in an Einstein Ring (B) and then processed to form first a convoluted image (C) and then deconvolved (D). Images are a work product from NASA's JPL, courtesy of V. Toth and S. Turyshev.

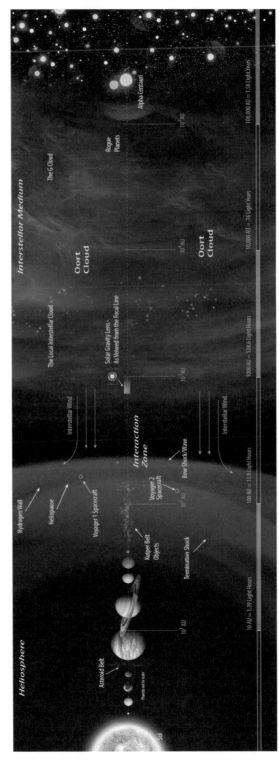

FIGURE 8 The interstellar medium from the Sun to the nearest star, Alpha Centauri. Distances are on a logarithmic scale. Courtesy of the Keck Institute for Space Studies.

much space out there to think they will have much success. Gravitationally detecting the planet at high precision is more likely. The best way to do it would be with a spacecraft with a precise navigation system and minimal perturbations from its own propulsion and heating systems and other nongravitational forces. Tracking such a spacecraft for a long time would reveal gravitational anomalies in the solar system. If we do a mission to the SGL focus, dropping off a small probe with a beacon and tracking it in the interstellar medium might be the way to search for Planet X.

According to some definitions, the heliopause is the boundary to interstellar space, although gravitationally the Sun remains dominant for a lot farther, beyond 100,000 AU. As noted in the last chapter, I consider interstellar travel a bridge too far, but its lure is compelling. Perhaps even more compelling is the goal of the discovery and investigation of life on other worlds. The solar gravity lens may provide our best chance to see such life, or at least evidence of it. A mission to the lens focus would be the appropriate next step once a likely life-bearing planet is identified. It is a challenging mission to where no one has gone before (or even yet proposed going to), but still it is thousands of times closer and less challenging than the interstellar flight goal. It can also be targeted to the primary exoplanet(s) of interest, not just the nearest one. This is truly advantageous—even more so when I show that we can enable the exploration of many exoplanets.

As mentioned above, we will need nuclear power far from the Sun, like that on the Voyager and New Horizons missions. Radioisotope thermoelectric generators (converting the heat of radioactivity into power) are adequate to operate spacecraft instruments and maneuver in the gravity lens focal region, but they are not enough for the propulsion to reach the speeds necessary for solar system escape. For that we would need a nuclear reactor powerful enough to supply a nuclear electric or nuclear thermal engine. Nuclear propulsion is one option for reaching the necessary speed to get to the SGL in a reasonable mission time, but it is expensive and not currently used because of environmental and safety concerns. Even if the political objections are overcome, it will require a large and expensive spacecraft. Conventional chemical rockets could be used if the spacecraft can use them very close to the Sun (for example, just 2–4 radii away). The enormous heat there would also require a heavy heat shield for the spacecraft—making that spacecraft and its launch vehicle very large and expensive. Even then, the speeds we can attain with chemical propulsion, no matter how big the rocket, are less than the desired 20 AU per year.

When people think of getting out of the solar system with a high-energy, high-speed spacecraft, they think of the big, powerful, and expensive solutions of nuclear propulsion and large rockets. An alternative is to think small and use the newly developed microsat capability for spacecraft (with a weight of less than one hundred kilograms) along with a solar sail capable of collecting enough solar power to place the microsat onto the hyperbolic trajectory. That way we avoid having to carry heavy, expendable propellant and instead get our propulsion externally from sunlight. The solar sail would get most of its thrust close to the Sun due to the increased power of sunlight there. The resulting boost in velocity (Delta-v) would increase the energy of the orbit, just as the Oberth maneuver (cited in chapter 6) does for the chemical rockets, and accelerate the spacecraft onto an escape trajectory. The perihelion required is small, but it is not so small as that required for chemical or electric rockets: instead of 2–4 solar radii passages, 10–15 radii should be sufficient. And the smaller spacecraft will be much easier to protect from the heat. I mentioned solar sailing earlier when discussing the laser sail idea for interstellar flight—but unlike that one, which requires the huge high-power laser array, solar sailing requires only sunlight. A smallsat–solar sail mission can reach a high solar system exit velocity and get to the SGL focus at modest cost and without a huge launch vehicle. It could even begin its flight in Earth orbit—negating the need for a large launch vehicle, using the sail to get out to interplanetary space and then spiral in toward the Sun.[7]

Solar sailing is an old idea—nineteenth-century scientists and mathematicians noted the force of sunlight, and Russian space pioneers Fridrikh Tsander and Konstantin Tsiolkovsky theorized about propelling spacecraft with the force of solar power. The first serious mission proposal was for a rendezvous with Halley's comet during its 1986 passage through the inner solar system. That was an idea before its time technologically and because the U.S. had no launch vehicle capable of carrying and deploying the sail. Working with Russian space engineers, The Planetary Society built the first solar sail spacecraft (Cosmos 1), but it was doomed when the Russian rocket (a submarine-launched former Soviet intermediate-range ballistic missile) failed to get off Earth in 2005. The first successful solar sail mission was IKAROS,[8] a Japanese space agency mission launched in 2010, which flew from Earth to the distance of Venus. In 2019, The Planetary Society launched LightSail 2, which lasted in orbit for three years. NASA is just now beginning to embrace solar sails, with a mission called Near-Earth Asteroid Scout, which was launched on the first mission of the new Space Launch System but never heard from as of late 2022.

The solar sail is propelled by reflected sunlight. The force comes from the transfer of momentum from the solar photons as they bounce off the highly reflective sail.[9] Just like on a terrestrial sailboat, the sail can be pointed so that the spacecraft can tack either toward the Sun (inward in the solar system) or away (outward), except here it is not the wind but sunlight pushing the sail. A terrestrial sailboat operates at the interface of wind and water, with its rudder setting the direction. The solar sailcraft operates at the interface of sunlight and gravity, the latter resulting in orbital motion around the Sun. If the sail is tilted so that the force adds to the orbit velocity, the spacecraft goes outward (the orbit gets bigger). If it is subtracted from the velocity vector, the spacecraft falls inward into a smaller orbit. That is how the spacecraft can be steered and how it can be made to move inwards or outwards through the solar system. As mentioned earlier, if the force is big enough the orbit gets so large that it opens out into a hyperbola on which the spacecraft can fly indefinitely out of the solar system. The bigger the sail, the smaller the mass, and the closer we can get to the Sun, the larger the force—potentially large enough for solar system escape. The larger that force, the faster the spacecraft will go on its exit hyperbola. Once on the exit trajectory moving away from the Sun, the force drops fast (by the square of the distance), and beyond Mars there is very little power from the Sun. So, it is important to use a sail such that we can get it as close to the Sun as possible. This, coupled with the need for a large sail area and a lightweight spacecraft mass specifies the spacecraft design parameters to get the largest possible velocity increase to fly through and out of the solar system.

The mission concept is depicted in figure 9: the mission launches on its heliocentric trajectory with minimum launch energy.[10] The sail tacks inward toward the Sun (by decreasing the orbital velocity) until it reaches its closest approach to the Sun (perihelion). At that point, the sail is reoriented to pick up maximum solar pressure and increase the velocity. The force is proportional to the sail area divided by the spacecraft mass, and is larger when the perihelion distance is smaller. The large area of the sail reflects as much sunlight as possible; the small mass of the spacecraft allows the momentum transfer (energy) to give a big velocity increase; and the small perihelion means more solar power hitting the sail. The sail should be as large as possible and the spacecraft mass as small as possible. The sail should also be thin and lightweight so that it won't take up much of the payload mass.

How large can we make the sail? At JPL in the mid-1970s, we proposed to build a sail that was fifteen kilometers in diameter (!) called a heliogyro. It was equivalent in size to a square sail that was a half mile on each side. That was far too audacious. The problem isn't just the size, it is whether the gossamer-like structure can keep its integrity and shape over years of flight, and whether it can be reasonably packaged and deployed however it is launched. Based on experiences now building sailcraft and deploying and controlling things in space, it seems that ten thousand square meters may be as large as we should think about at present. That would be a 100 m × 100 m sail, which might require boom stiffening or guy wires. A novel sailcraft design called SunVane is now being studied that uses multiple sails (or vanes) that are smaller, easier to package, and more controllable.[11]

How light can we make the spacecraft? There are two parts to that—how thin the sail material can be and how small the spacecraft can be. While there are nanosats of less than ten kilograms in mass, they are not yet robust enough for many years of interplanetary flight. But even when they are, to work in the outer solar system and interstellar medium, they will need to have a power source independent of the Sun—that is, nuclear power. Even

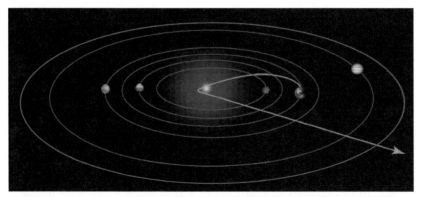

FIGURE 9 Computer art of the fast solar sail trajectory launching from Earth, flying around the Sun, and then through the solar system. The image is a JPL work product, courtesy of Artur Davoyan.

using advanced nuclear batteries or the decay heat of long half-life radioactive elements in a generator, we will probably be limited to spacecraft that weigh tens of kilograms—perhaps fifty kilograms, especially if the craft carries good science instrumentation. The sail material might be as thin as one micron, with a density of less than one gram per square meter. A sail of ten thousand square meters would then weigh about ten kilograms. That still leaves forty kilograms for the spacecraft, science instruments, and power—not possible for today's technology, but probably possible in the not-too-distant future. A ten-thousand-square-meter sail on a fifty-kilogram spacecraft has an area-to-mass ratio of two hundred square meters per kilogram.

How close can we go to the Sun? This also will be determined by the sail and spacecraft separately. For the latter, we will have to protect all the electronic components with shielding. That will weigh a lot if we get too close to the Sun. But of even more concern is the possible melting of the sail material. If the sail has a plastic substrate (for example, Kapton), it is probably limited to a distance of about 0.2 AU (about thirty solar radii), but studies of pure aluminum and some new exotic materials (called metamaterials[12]) suggest that it may be possible to go to ten solar radii or even less with a solar sail.

Current technology limits all three of these factors—how close we can get to the Sun, how big we can make the sail, and how small we can make the spacecraft. But what is especially encouraging about the new architecture of smallsats and solar sails for fast transit through and out of the solar system is that these are technologies now undergoing rapid development with active research. They are not the old way of just building bigger rockets with more propulsion, but a new way using miniaturization and advanced materials. Developments in electronics and other areas of physics are helping to drive technological developments for space applications.

So, while reaching 20 AU per year (about six times faster than the fastest escape velocity ever achieved) is currently a challenge, advanced technology may one day permit sails the size of a football field and spacecraft the size of a modern CubeSat, composed of exotic materials of high reflectivity and able to withstand the very high temperatures close to the Sun. Such technology might permit going twice as fast, 40 AU per year or higher. Speed is important, not just for getting out there quickly, but also for dealing with the years of collecting, integrating, and processing data along the focal line once the imaging begins. Just as we do with planetary orbiters, we will want our focal line mission to extend for years to enable continuous observations of an exoplanet while it orbits its parent star.[13] That is a long-term science objective, although the payoff, imaging the surface of an exoplanet at kilometer-scale resolution, would be uniquely important. If we can do it, it will be worth waiting for.

Two missions being proposed to fly through and out of the solar system should be mentioned, even though they are not astrobiological missions. The first is called Interstellar Probe, with a plan to reach the interstellar medium beyond 200 AU to measure the fields and particles environment of this pristine interstellar wind. Proposed by the Johns Hopkins University Applied Physics Lab, it would use an Oberth maneuver with a low solar perihelion. The second proposed mission, by the Chinese National Space Administration, is called Interstellar Express—with the goal of reaching 100 AU on the one-hundredth anniversary of the Chinese Communist Party takeover in China, that is, in 2049. It is planning multiple gravity assists of Earth in the inner solar system and then by Jupiter in order to reach sufficient velocity. Naming them "interstellar" is a bit of a stretch since neither has an ambition even as far as an SGL focus, but it testifies to the lure of interstellar travel.

Perhaps the most important propulsion technology isn't propulsion at all—it is the mechanical and electronic miniaturization that has changed everything in our daily lives. Phones on our wrists, computers in our pockets, cameras in our eyeglasses (and in every nook and cranny around us) are a few examples. The decreased size of things makes them convenient; but more importantly, it makes them cheap (affordable). That is because they can be mass produced. One day it may be possible to mass produce smallsats, too, at low cost. In fact, this has already started to happen—for example, hundreds of schools are now building spacecraft, often Cubesats, though there are other kinds. Commercial companies are putting up networks of communications and internet satellites—not in quantities of six to twelve, as done a few years ago, but by the hundreds and thousands. With science missions, it is a different story. Rarely can you get two spacecraft approved at the same time—as was done for the original Mariner, Viking, and Voyager missions. Partly because of scientists' insatiable appetite for improving things, and partly because of the rarity of launch opportunities, new exploration missions tend to be bigger than their predecessors. Instead of exploring the Martian surface with dozens of identical rovers, new ones launched every Martian year, we instead are going back to doing bigger, unique missions once per decade or so.

The smallsat approach, which I introduced here as a component of the solar sailcraft for obtaining high speed, will enable more places to be explored quickly, and if sent on different missions to different exoplanet foci of the SGL they will also enable multiple exoplanets to be explored concurrently. We need that—we can't expect to explore life on other worlds with one or two lucky guesses per century. We need multiple long-lived spacecraft for exoplanet exploration just as we need many orbiters to explore here in our solar system.

With long mission times, and likely with target exoplanets in several different star systems, we will need a low-cost, highly repeatable, and flexible spacecraft architecture—one that might permit a series of small missions rather than one with a traditional large, complex spacecraft. Its velocity might also be boosted with a hybrid approach, adding electric propulsion to the solar sail. We need to carry a small amount of electric propulsion anyway to have maneuvering capability near the SGL focus. Such a multiple smallsat–sailcraft, close-to-the-Sun perihelion mission design is a new approach to space exploration. The multiple-spacecraft approach also permits redundancy for increased reliability. Of course, all exploration involves multiple spacecraft—just as ocean exploration involved multiple ships and voyages. The exploration of our solar system has been carried out by many spacecraft—dozens at the Moon, Mars, and Venus, and several at Mercury, Jupiter, and Saturn, with more planned. Similarly, multiple telescope missions observe the universe. But going to the far reaches of the solar system and into the interstellar medium takes much longer times. In particular, observing exoplanets with the SGL not only would take a long time, but there are many different planets we might want to observe. The small-spacecraft approach allows that exploration to be carried out simultaneously, rather than sequentially waiting for one mission to end before starting another—like the Niña, Pinta, and Santa Maria, the multiple ship approach provides increased reliability and robustness to exploration. If we employ this new approach, and do it right, we might be able to explore dozens of potentially habitable (or even inhabited) exoplanets in the second half of this century with high-resolution, detailed observations and direct imaging. Right now, with today's technology, it would probably take several dozens of years to reach the SGL focus and get images of an exoplanet—but with smaller spacecraft and advanced sail materials to go close to the Sun, we might be able to reach speeds greater than 40 or even 50 AU per year and send multiple spacecraft to observe multiple exoplanets, all for less than a cost of a single space telescope (JWST, for example). In the next chapter, I offer series of interstellar precursors that lead us to 1,000 AU, with the ability to image potentially inhabited exoplanets—wherever they might be.

The solar sail–smallsat mission design is low cost and permits multiple spacecraft for the study of multiple exoplanets. But it is certainly not the only propulsion option. Different kinds of sails are being studied—a refractive sail instead of a reflective sail permits a larger thrust for a given area and mass of the spacecraft, and an electromagnetic sail derives its propulsion from interplanetary electrons and protons instead of light. Both are only theoretical at this point. I cited the limits of chemical propulsion and solar electric propulsion, and the cost, size, and political difficulties with nuclear electric and nuclear thermal propulsion. One other concept, long dreamed of, is being studied: nuclear fusion, small enough to fit into a single launch vehicle and provide solar system exit velocities of more than 50 AU per year. Nuclear fusion, according to one of its proponents, would change everything about the possibilities of exploring beyond the solar system.[14] That is true, but it has been said for decades, and we don't even have nuclear fusion supplying power for any application on Earth. The space limitations are bigger and much costlier. One doesn't want to say it will never happen, but practical approaches to reach the SGL focus, or to do anything at speeds greater than 20 AU per year (approximately one hundred kilometers per second), in the foreseeable future require thinking small—smallsats with solar sails.

8

REAL INTERSTELLAR EXPLORATION

The bridge too far of interstellar flight can be crossed virtually, making real exploration possible.

HAVE NOW SET the stage to propose "real" interstellar exploration—not that of science fiction or that which is theoretically possible (in physics) but realistically impractical. "Real" is put in quotes here because, in the context of exploring areas we can't reach (in any reasonable lifetime), it means the use of virtual reality (VR). "Real" meaning "virtual" may sound like an oxymoron, but not here: the VR that we can create would use real data and scientific methods to process and analyze that data. We can get data using Nature's telescope, the solar gravity lens, to image exoplanets identified from Earth-based and Earth-orbit telescopes as the most promising to harbor life. It also would use modern information technology to create immersive and participatory methods for scientists to explore the data—with the same definition of exploration I used at the beginning of this book: an opportunity for adventure and discovery. The ability to observe multiple interesting exoplanets for long times, with high-resolution imaging and spectroscopy with one hundred billion times magnification, and then immerse oneself in those observations is "real" exploration. VR with real data should allow us to use all our senses to experience the conditions on exoplanets—maybe not instantly, but a lot more quickly than we could ever get to one.

The distinction between "real" and "virtual" exploration is not easy to characterize. We might call having a human walk on Mars "real" exploration, in contrast to having data from a robotic spacecraft sent back to Earth for analysis. But a human on Mars, by necessity, would be encased in a space suit, using a myriad of sensors including cameras as much as they use their own eyes. And they would no more feel or smell or taste or hear than a robot would—all require external sensors. Furthermore, a human on Mars would be extremely time and safety constrained, greatly limiting their ability to process information. With a huge amount of robotically collected data sent into VR models, a virtual explorer on Earth could methodically roam through it all in three dimensions, going into hard-to-reach and dangerous places (such as climbing down a canyon cliff). Who is the real explorer?[1] Referring to the simple definition I gave for exploration in the introduction (discovery plus adventure), there is no doubt that a human on Mars in a spacesuit would have an adventure and make discoveries, but so would a human operating an advanced robot on Mars in VR. One might have more adventure than the other, while the other might make more discoveries.

If the only type of data we get is pictures, as we would get from SGL imaging, it might be argued this doesn't make for real exploration. That requires some immersion of the observer into that which is being observed. VR permits that, but thus far (for us consumers used to seeing it in games or in entertainment centers), it has been very limited. For now, VR users still look at it on a two-dimensional display and can feel disoriented. And it is still computationally demanding, which imposes severe limits on speed and the quantity of data that can be used. But this is a very active subject of research, and, as with the microminiaturization of small spacecraft, space scientists can take advantage of larger research areas in the context of consumer electronics and medical computation systems. An AI technique called deep learning has been applied to VR models through a process called tensor holography, which permits the construction of 3D holograms that then can be explored in a very real fashion with the explorer immersed in the scene.[2] Given the advances happening right now in the use of remote observation data for exploration, it is not hard to imagine scientists moving around on the surfaces of planets "virtually" while exploring the real images of those worlds in a 3D hologram as if they were there—perhaps even tripping over rocks or fissures on the surface.[3]

While this is all a bit "way out there," it is certainly less speculative than the idea of sending people or even probes over the bridge too far to exoplanets. Bringing exoplanets to us makes it possible to conduct real interstellar exploration in a modern way—with remotely sensed data processed with AI and VR tools. Studying many exoplanets in this way would enable us to discover life and/or learn about the conditions for life in the universe. Before the Space Age, there was no field of comparative planetology. Now it is routine and fundamental to compare data from many planets to understand the nuances and complexities of the dynamic processes on each of them—and on Earth. Someday, given the many exoplanets that likely have forms of life or conditions relevant to life, we might develop a new field of comparative life studies, perhaps calling it (for the scientists) comparative astrobiology. That is pretty exciting, and that is why the title of this book says we are "not lonely."

What exoplanets to explore? Where to begin the study of comparative astrobiology? It is clearly premature to select targets of interest now—we are only in the earliest stages of discovery and identification of exoplanets, limited to observing the shorter-period ones around cooler stars and the larger ones around Sun-like stars. I cited a number of newer-generation telescopes planned for the 2020s and 2030s that will focus more on exoplanets, and we can expect the discovery of a lot more life signatures and conditions that hint at habitability. We won't observe surface features until we get the SGL focus mission, but we will get a lot more data that will help select candidate targets for such a mission(s). In particular we will get spectral data, which even from one-pixel observations will be useful. Similar to the repeated "Water found on Mars!" headlines a few decades ago, we will undoubtedly get headlines proclaiming hints of life on other worlds. But really, don't believe it until we see it. We have been tricked a couple of times in recent years (by methane discoveries on Mars and phosphene discoveries on Venus) into making false claims about life detection. The science data was good, but leaps of interpretation were sometimes made. Hints of life are not life.

To enable our exploration of these worlds, we must carry out multiple missions to the individual SGL focal lines of multiple exoplanets. Space missions to the SGL focal region (600–1,000 AU along the focal line) are, as described in the previous chapter, demanding. But the technologies are promising—the miniaturization of spacecraft with increased capabilities is rapidly advancing, and new metamaterials and carbon-carbon materials including graphene are being invented and fabricated. They will enable really fast small solar sailcraft. We also have a single network of thousands of satellites being developed in Earth orbit. It is not hard to envision smaller networks of spacecraft, each of low mass, being able to travel to the SGL focus on the type of trajectory described in the previous chapter at speeds greater than 20 AU per year. Nor is it hard to extrapolate doing this to several exoplanet focal lines.

Creating the interplanetary smallsat–sailcraft architecture is not a one-step process; space missions require development, with steps increasing in complexity and capability. For fast space flight deep into the interstellar medium, technology requirements include small nuclear batteries to operate microthrusters and a laser or radio transmitter and instruments on the smallsat, along with advanced image processing of the exoplanet pixels observed in the Einstein Ring. "Image processing" might include those new subjects mentioned above—deep learning, tensor holography, and whatever else is necessary to create, display, and then interact with the VR model of a real exoplanet world.

Solar sail technologies were also cited earlier—carbon-carbon composites and metamaterials—and their use may permit the doubling of the speed through the solar system, to 40 or even 50 AU per year. I noted Prof. Artur Davoyan's work on Extreme Solar Sails (meaning materials that can withstand high temperatures near the Sun) in a NASA study of metamaterials. Other NASA Innovative Advanced Concepts Studies are considering different types of solar sails—not reflective but refractive, and not optical but electromagnetic.

There are also alternatives to the use of solar sails—I earlier cited proposed nuclear propulsion schemes (nuclear electric and nuclear thermal), but they require on-board reactors that come with large fiscal, environmental, and political requirements. Nuclear propulsion would also require large spacecraft with big rockets to launch them. The ability to afford multiple spacecraft to image multiple targets would be very small. The mission design described in the last chapter for getting through the solar system at a speed faster than 20 AU per year with small sailcraft offers a new paradigm for access to deep space with small, fast-flying, low-cost rockets.

To summarize: the development of an SGL focus mission able to image many candidate habitable or inhabited planets requires a new approach to deep space mission design. The key innovative ideas are:

- very lightweight interplanetary smallsats
- a new high area-to-mass ratio design for solar sailcraft
- the deconvolution of images created by a pixel-by-pixel sampling of Einstein Rings over tens of years and hundreds of AU
- very lightweight nuclear power (a battery) for the spacecraft and its communications system over distances of 600–900 AU
- the creation of a VR model to simulate in-situ exploration of imaged exoplanets
- the use of multiple smallsat–sailcraft for multiple exoplanets that can accommodate the telescope spectrometer, the communications system, and sufficient power and microthruster propellant to carry out imaging operations.

This is a lot to swallow—too big a bite for a single gulp. However, unlike interstellar flight to exoplanets, it is not a bridge too far. An incremental approach to prove each of these innovations is possible, with each step increasing mission capability and reliability. The incremental approach dovetails well with a step-by-step approach of deeper space science missions, reaching distant interesting targets once thought to require impractically long mission times. Each of the steps can then yield technical and scientific benefits on the way to the grand goal of imaging exoplanets.

The first step is to prove the new paradigm for access to, and then through, the solar system. This can be done with a very simple spacecraft with sailcraft technology and a low perihelion flight around the Sun to achieve a record-breaking speed. This first step would merely demonstrate the technology, with no need for a payload of ambitious science instruments. Breaking the speed record would capture the public imagination, signalling the start of a new era of fast interplanetary and interstellar exploration. This mission might carry very small (of order one kilogram) technology experiments and cultural payloads akin to those of Voyager and Pioneer. Perhaps an interstellar message might be carried, since even this mission would exit the solar system at about twice the speed of Voyager.

(The subject of interstellar messaging is discussed in appendix B, which traces its history and significance.)

Let's call this first step the LightCraft Demonstration Flight (LDF). It should be very low cost (of order millions of dollars), perhaps even privately funded. Succeeding steps will add scientific payload and test other technologies for advanced exploration hundreds of AU away. With our focus on "real" interstellar exploration, we might consider following up the LDF with a mission to rendezvous with an ISO. Such a mission is currently challenging, because we have no spacecraft capable of achieving ISO speeds, nor can we launch one as soon as an ISO is discovered entering our solar system. However, this would be possible with a smallsat–sailcraft.

I mentioned ISOs when I discussed the first one discovered, the strange looking 'Oumuamua (figure 6, p. 67). Its discovery in 2017 was followed by another one just two years later, Borisov. It was more conventional, looking and behaving like a comet from interstellar space (as its discovery article was titled in the scientific literature). ISOs can't be predicted (at least not far in advance) because they are very small (like comets and asteroids) and come from very far away on a fast hyperbolic trajectory for a single pass through the solar system. 'Oumuamua reached a perihelion of 0.25 AU, with a hyperbolic velocity of about 5 AU per year, while Borisov reached only about 2 AU (farther out than Mars) with a hyperbolic velocity of about 6 AU per year.[4] We can guess that many more ISOs will be discovered in the future: astronomers predict perhaps about one per year (although none have been discovered since 2019). Getting up close to and exploring an ISO would be interesting and important—a closeup look of something from another star system. Scientists are interested in ISOs—the European Space Agency and NASA are considering several concepts for fast flybys of potential newly discovered ones. Conventional approaches involve long flight times, with flybys occurring far out in the solar system, and large rockets to launch them from Earth on short notice.

If we could actually rendezvous with and land on one, we could bring back samples to analyze in Earth-based laboratories, or we could "ride" on it to interstellar space (as Joseph Breeden hypothesized in the work cited in chapter 6). But a mission to an ISO requires (1) an extremely fast reaction time to a discovery and (2) a spacecraft capable of going faster than the ISO. The latter can be done by a smallsat–solar sail and the former can be achieved by the mission design described earlier whereby the spacecraft is launched from Earth orbit inward toward and then around the Sun to achieve a high solar system exit speed (figure 10). Instead of waiting until an ISO is discovered, we can launch from Earth into a low solar orbit and just wait there until the discovery. The time wouldn't be wasted— important measurements of the solar environment and monitoring of the Sun could be taken and used by scientists. When an ISO is discovered and its orbit determined, we can direct our sailcraft to then go around the Sun to catch and rendezvous with the ISO. We can do this no matter what the ISO's inclination is, since the solar sail can be used to continuously raise the inclination, even flipping the orbit over to make it retrograde if required.[5] While an ambitious and first-of-its-kind mission, it is much less demanding than an SGL focus mission, having only to fly for a few years at speeds less than 8 AU per year with a smallsat and a reasonably sized solar sail. The mission is possible with current technology—for example, in the inner solar system we can use solar power with photovoltaic cells on the sail and also use the sail for high data rate communication. Yet it would also be an important technological step forward toward an SGL focus mission, advancing such systems as power, control, communications, and navigation. The science payoff would be high.

The lightsail–smallsat approach described here is much lower cost and much more responsive to discovery than conventional planetary missions. It could even enable a rendezvous or a possible sample return. We could place a solar sail spacecraft in a low circular orbit around the Sun, inside Mercury's orbit, say at a distance of 0.2–0.25 AU. We could do this with the same mission concept described in the last chapter—launch from Earth with a small rocket or as a secondary payload on a larger mission and spiral in toward the Sun by tacking the sail continuously to reduce orbital velocity. At any point on the spiral, if we stop the propulsion (for example, by turning the sail edge-on to the Sun) we shift to a circular orbit around the Sun. We can consider this a parking orbit in which to wait for an ISO discovery, and we can utilize the time in this orbit for scientific study of the Sun and the inner solar system environment. A Solar Polar Orbiter provides a mission of high interest to scientists who have no other economical way to reach the ecliptic pole to view the poles of the Sun. Secondary targets among the many comets and asteroids in the solar system can be selected as a backup in the case of no early discovery of an ISO. Once a target is discovered or selected, the sail can be pointed to modify the orbit and direct the spacecraft to the target either as an intercept mission or to enable a rendezvous (matching the object's position and speed). This provides a fast and flexible response to new discoveries and observations.

The reconnaissance and survey of the distant small bodies of the solar system and beyond particularly need the capabilities of fast flight, multiple spacecraft, and smallsats. These objects include long-period comets, ISOs passing through the solar system,[6] Kuiper belt objects (KBOs), and the many moons of the outer planets. Studying their nature is an important component of the field of planetary science. Exploring their comparative planetology, even with fast flybys, could serve as an excellent stepping-stone to missions heading out of the solar system.

Similar to KBOs are long-period comets. They are on elliptical orbits, with aphelions in or near the Oort cloud. The Oort cloud is where all comets in our solar system originate, a conglomeration of perhaps one trillion comets between 25,000 and one million AU from Earth. The objects at that distance are just dirty snowballs (icy pieces of space debris)—that far out, they don't look like comets. Comets only develop heads and tails

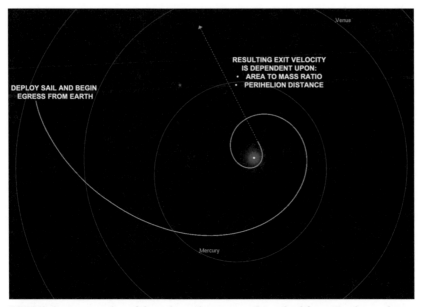

FIGURE 10 Computer plot of the solar sail trajectory launched from Earth to achieve a 20 AU per year exit velocity from the solar system. Image by Darren Garber, NXTRAC.

when they approach the inner solar system and they are heated up by the Sun, boiling off the volatiles that then get caught in the solar wind. Comets in the Oort cloud just hang out there until an occasional one is perturbed enough by some gravitational source that its orbital velocity decreases and it journeys inward. The periods of these comets are hundreds to tens of thousands of years. One such comet has now been identified. C/2014 UN271 (Bernardinelli–Bernstein), or simply 2014 UN271, is a large Oort cloud comet with an aphelion of approximately 50,000 AU and a perihelion of approximately 11 AU. It will reach perihelion in January 2031 with a speed of between 2 and 2.5 AU per year. That would be an interesting comet to get to and observe. We can do that with a smallsat–solar sail. The best-known comet is Comet Halley. But at seventy-six years, its period is really not that long—which is why it has been observed many times in human history. The next time it comes into the inner solar system is in the year 2061. The previous time was 1986, and while I was at JPL we put together a mission concept to rendezvous with it with a solar sail spacecraft. But that was before smallsats, and our ideas were just too audacious for the time. The Soviet Union, the European Space Agency, and Japan all mounted missions to intercept Halley's comet (sadly, the United States did not).[7] They were fast flybys—the first ever to a comet. Hopefully, in 2061, we'll be ready with an armada of smallsats and solar sails to rendezvous with the comet and study it up close.

Getting to and operating in the Kuiper belt to study several KBOs would take us almost into the interstellar medium. (The Kuiper belt extends from Pluto at about 40 AU to perhaps 120 AU, whereas the interstellar medium begins at around 140 AU.) Distances out there are immense, and a survey and study of KBOs would require many spacecraft. This is another problem that the smallsat–solar sail design can solve—the small spacecraft will be relatively inexpensive, and multiple launches from Earth would not require big and expensive launch vehicles. While the flybys of KBOs would be quick (say, at twenty-five kilometers per second), passing by them in just minutes, discoveries could be followed up with other reconnaissance spacecraft, and the comparative data from many KBO flybys would be very valuable. The ISO and KBO interstellar precursor missions could be carried out in the late 2020s and early 2030s, enabling technology readiness and the development of the SGL focus mission to be done by the middle 2030s, with the potential of obtaining images of habitable or inhabited exoplanets by the 2060s and 2070s. It would also enable a Comet Halley rendezvous during its 2061 apparition.

Whether or not interstellar travel remains a bridge too far, the ability to explore many habitable and interesting exoplanets in the latter quarters of this century will certainly open up a new age of science and exploration, and a new understanding about the possibilities of extraterrestrial life. Just as our notion of human spaceflight has to change (as described in an earlier book[8]) to take into account both evolving technology and evolving human life, we must change our notion of interstellar flight from the science fiction driven notion of powerful spaceships travelling to other star systems at relativistic (or even faster) speeds to one of interstellar exploration by remotely looking at the plethora of habitable planets in our Galaxy through Nature's telescope. We can do this systematically with small spacecraft, small enough to be affordable so that we (all space-faring nations) can build dozens, or even hundreds, of them and robustly target exoplanets, ISOs, Oort cloud comets, and KBOs just as we do now with spacecraft in our "little" solar system of eight planets and many small bodies. This can begin in the latter half of the twenty-first century and can keep us busy for at least a couple more centuries after that. With all the data we get from years of observing exoplanets through the solar gravity lens we can start building VR models to enable planetary scientists and the general public to truly explore new worlds. And through study of those worlds that do have life—or even some precursors of life in their chemical compositions—we can begin to understand our own life here on Earth.

What I have just outlined in the last five paragraphs would be a real program to explore extraterrestrial life and interstellar worlds. Creating a production line of smallsat–sailcraft would create the means to step through the solar system into the interstellar medium and explore dozens, or even hundreds, of worlds—including habitable exoplanets. This approach to *real* interstellar exploration is shown in figure 11. The sequence shown there might be considered as stepping into the interstellar medium ultimately to explore habitable (and maybe inhabited) exoplanets:

- 2025–2026: the LightCraft Demonstration Flight with a one-to-two-year mission time, followed by a Solar Polar Orbiter
- 2028 . . . : mission(s) to explore newly discovered ISOs; fast Europa and Enceladus plume fly-throughs

- 2030–2040: Uranus, Neptune, and KBO mission(s), interstellar medium probes
- 2035–2040 . . . : a series of missions to SGL foci of exoplanets

This last item is what enables real interstellar exploration—missions observing many worlds around other star systems, most likely showing signs of life. When we bring that data to Earth, reconstruct it in our VR models, display it with our AI holograms, and enable scientists and possibly even tourists to visit those displays, then in a figurative sense we will be interstellar explorers interacting with life in our Galaxy.

This isn't the way Gene Roddenberry envisioned our "going where no one has gone before" when he conceived and developed *Star Trek*. I was privileged to know Gene through The Planetary Society. He helped us in our earliest days reach out to his millions of fans, explaining that the scientific advancement of solar system exploration advocated by Carl Sagan, Bruce Murray, and myself was strongly related to the multiplanet species vision of *Star Trek*.[9] *Star Trek* wasn't scientific, but it was foundational and inspired science, and was revolutionary in many ways, including in the casting of the original communications officer, Lieutenant Uhuru, played by Nichelle Nichols—a wonderful person who became a good friend of ours at the Society.[10] As is clearly evident in this book, I don't think

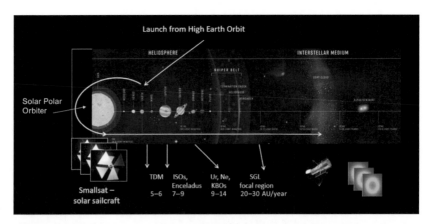

FIGURE 11 A sequence of ever-faster missions into and then through the solar system: the LightCraft technology demonstration mission (TDM) and science missions in our solar system to ISOs, Uranus, Neptune, etc., leading to the exploration of exoplanets with the solar gravity lens. Image by the author.

humankind is going to be wandering around the Galaxy, interacting with interesting intelligent species—we are alone. But our exploration of other worlds in our own solar system, and then of many other worlds and most likely other life in other star systems with virtual, real interstellar exploration, will certainly prevent us from feeling lonely. We go to space to find ourselves, as T. S. Eliot said:[11]

We shall not cease from exploration
And the end of all our exploring
Will be to arrive where we started
And know the place for the first time.

The poem continues with speculation about the last Earth "left to discover" and its origin—"that which was the beginning" (at least that is how I read it). It seems not unreasonable to interpret this as a reference to humankind's looking outward to discover ourselves. I believe we will learn about ourselves as we explore life on other worlds and not mind at all that we are alone, but not lonely.

9

COMPARATIVE ASTROBIOLOGY

*The future holds the promise of discovering and studying life
on many worlds. This will lead to a new field of comparative
astrobiology, the results of which are unpredictable but certain
to be far-reaching.*

IN CHAPTER 5, I noted that exploration in our solar system is already
uncovering several worlds of astrobiological interest—Mars, Europa, and
Enceladus being the top three and others being more longshots (figure 12).
Multiplying our solar system experience by the increasing number of potentially habitable exoplanets that we can hope to explore in many ways—from
telescopes on Earth and new ones in space and, in a decade or two, through
multiple missions to their SGL foci—we might anticipate life discoveries
on dozens of worlds in the next fifty years or so. It's time to think about that.

The discovery of life on another world would be (to use a pun) earth-
shattering in its implications. Perhaps the 1543 publication by Copernicus
of *De revolutionibus orbium coelestium* (On the Revolution of the Heavenly
Spheres) gives us the best example. No one doubts the profound implications
of understanding that the planets revolve about the Sun, as he described, versus the previous view that they (and the Sun) rotated around Earth. Yet, daily
life was not much affected by this discovery. Europe continued its wars and
religious upheavals, and began the Age of Exploration without the need for
heliocentric calculations. Change happens faster now, but still the impact of
a discovery of life on another world won't cause much immediate change in
human behavior. Its impact, however, will be as profound as that of the heliocentric theory of orbits, and we can anticipate the development of whole
new ways of understanding the universe and our place in it. It will not be a

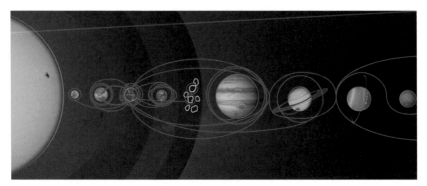

FIGURE 12 Artist's depiction of NASA's astrobiology missions—many candidate worlds in our own solar system. Note that only Earth is in the Sun's habitable zone—the area of focus for exoplanet scientists. Yet in our own solar system we have more than a half dozen potentially habitable worlds of interest. Courtesy of NASA.

simple "yes there is life" or "no there is not" discovery; rather, it will yield lots of details about the physical, chemical, and biological aspects of life. It will answer all sorts of questions: big questions such as whether other life has DNA/RNA and is carbon based, little questions about the proportions of certain chemicals and molecules, and fundamental questions about how it propagates and evolves. Is it animal or vegetable or something else? A new field of study will be opened: comparative astrobiology—the examination and understanding of life on other worlds as compared to ours.

I already noted how little we know about life on Earth—how it started, why it evolved so slowly at first, what caused the Cambrian explosion, whether the evolutionary path to intelligence was an accident or an eventuality. Even more so, what will the future bring—stasis, speciation, extinction, replacement of the human species? Just as comparative planetology of the rocky bodies of the solar system has led to a greater environmental understanding of Earth (for example, of volcanism, the greenhouse effect, climate change, and plate tectonics), we can expect comparative astrobiology to help us understand those questions about life's evolution on Earth, as well as perhaps fill in the missing links between prebiotic chemistry and biology and simple single-celled life and its evolution into complexity (that is, us).

The biggest problem for astrobiologists might be recognizing life. This is not trivial, even if we were to stand right next to it; for example, remember the many months of debate about the putative life that may have been discovered in the Martian meteorite. It was convincing enough to warrant a peer-reviewed, published science paper, a NASA press release, and then a presidential announcement. But later studies led to the conclusion it was likely not life at all. It is still an uncertain conclusion. This was an extraterrestrial case—there are many discoveries on Earth or in laboratories that are as uncertain. The identification of life will be much harder and even more uncertain with remote discoveries—for example, looking from orbit around Mars or Europa, or even through the eyes of a lander. Remote observations of exoplanets are even harder. Without the SGL, all our observations of an exoplanet will be smaller than one pixel.

Before considering what to look for, let's first consider the necessary conditions for life: (1) an energy source, (2) a liquid solvent, and (3) nutrients. Each of these is something to look for, and finding a combination of them would be most promising. Life must also evolve (that is, it can't be static): when conditions change, it must change and adapt—that probably means complex molecules and molecular reactions interacting with the environment. What constraints this puts on environments for life is uncertain, but we know of some pretty extreme environments on Earth (cold, hot, dark, dry . . .) that support life and we suspect there are limits like those on the hellish surface of Venus or in the crushing deep atmospheres of the outer planets.

The ancient parable about a group of blind men trying to describe an elephant after touching different parts of it might apply here with different scientific disciplines trying to describe life from their different vantage points. A beautiful summary of this is given in table 3.[1]

TABLE 3 Disciplinary Perspectives on Signatures of Living Processes

SCIENTIFIC DISCIPLINE	TYPICAL MEASURES OF LIFE AND OBJECTS OF STUDY	BIOSIGNATURE RELEVANCE
Mathematics	Theorems, proofs, calculus, algebra, number theory, geometry, probability and statistics, computational science (Chaitin, 2012).	The language of science. Quantitative frameworks of relationships in nature.
Physics	Motion of mass and electromagnetic energy, quantum behavior, organization, dissipative structures, collective behavior, emergence, information, networks, molecular machines (Schrodinger, 1943; Goldenfeld and Woese, 2011; Walker, 2017)	Conservation laws to constrain abiotic context. Systems interactions of biological processes

TABLE 3 *continued*

SCIENTIFIC DISCIPLINE	TYPICAL MEASURES OF LIFE AND OBJECTS OF STUDY	BIOSIGNATURE RELEVANCE
Chemistry, biophysics	Redox potential, Gibbs free energy (Hoehler, 2007; Smith and Morowitz, 2016)	Metabolic processes that alter the redox state of the environment
Microbiology, molecular biology, biochemistry	Cells, genes, genomes, RNA, proteins, metabolism (Woese, 1998)	Constraints on evolutionary path requirements for a type of life to emerge. Metabolic products that can be strictly biogenic.
Geologists, geophysics	Isotope fractionation, morphology, fossils (Knoll, 2015)	Planet formation factors that determine prebiotic elements. Plate tectonics to allow a carbon cycle.
Philosophy	Emergence, meaning, goal-directedness (many reviewed in Mix, 2015)	Definitions of intelligence, optimality.
Ecology	Ecosystem, community dynamics, scaling laws, keystone species (May et al., 1974; Pikuta et al., 2007; Amaral-Zettler et al., 2011)	System interactions that lead to dominance or community mixes of particular kinds of life, determining what biosignatures will be detectable.
Biochemistry, geochemistry	Elemental cycling (Schlesinger and Bernhardt, 2013), serpentinization	Budgeting of the fluxes and stocks of particular molecules, wherein the net accumulated stock or phasing of fluxes may be detectable biosignatures.
Astronomy	Planetary-scale spectral signatures, molecular line lists, remote observation (Meadows, 2005; Seager, 2014; Seager and Bains, 2015; Seager et al., 2016)	Stellar context for life determines the radiative balance and elemental composition of a planet. Detection of biosignatures in planetary spectra from transits or direct imaging.

References cited: Linda A. Amaral-Zettler, Erik R. Zettler, Susanna M. Theroux, Carmen Palacios, Angeles Aguilera, and Ricardo Amils, "Microbial Community Structure Across the Tree of Life in the Extreme Río Tinto," *ISME Journal* 5 (2011):

42–50; Gregory Chaiten, *Proving Darwin: Making Biology Mathematical* (New York: Pantheon Books, 2012); Nigel Goldenfeld and Carl Woese, "Life Is Physics: Evolution as a Collective Phenomenon Far from Equilibrium," *Annual Review of Condensed Matter Physics* 2 (March 2011): 375–399; Tori M. Hoehler, "An Energy Balance Concept for Habitability," *Astrobiology* 7, no. 6 (December, 2007): 824–838; Andrew H. Knoll, *Life on a Young Planet: The First Three Billion Years of Evolution on Earth* (Princeton: Princeton Univ. Press, 2015); R. M. May, G. R. Conway, M. P. Hassell, and T. R. E. Southwood, "Time Delays, Density-dependence and Single-species Oscillations," *Journal of Animal Ecology* 43, no. 3 (October, 1974): 747–770; Victoria S. Meadows, "Modelling the Diversity of Extrasolar Terrestrial Planets," *Proceedings of the International Astronomical Union* 1, no. C200 (2005): 25–34; Lucas John Mix, "Defending Definitions of Life," *Astrobiology* 15, no. 1 (January, 2015): 15–19; E. V. Pikuta, R. B. Hoover, and J. Tang, "Microbial Extremophiles at the Limits of Life," *Critical Reviews in Microbiology* 33, no. 3 (January, 2007): 183–209; William H. Schlesinger and Emily S. Bernhardt, *Biogeochemistry: An Analysis of Global Change*, 3rd ed. (San Diego: Academic Press, 2013); Erwin Schrödinger, *What Is Life?* (Cambridge: Cambridge Univ. Press, 1944); Sara Seager, "The Future of Spectroscopic Life Detection on Exoplanets," *Proceedings of the National Academy of Sciences* 111, no. 35 (2014): 12634–12640; Sara Seager and William Bains, "The Search for Signs of Life on Exoplanets at the Interface of Chemistry and Planetary Science," *Science Advances* 1 (March 2015): e1500047; S. Seager, W. W. Bains, and J. J. Petkowski, "Toward a List of Molecules as Potential Biosignature Gases for the Search for Life on Exoplanets and Applications to Terrestrial Biochemisty," *Astrobiology* 16, no. 6 (June, 2016): 465–485; Eric Smith and Harold J. Morowitz, *The Origin and Nature of Life on Earth: The Emergence of the Fourth Geosphere* (Cambridge: Cambridge Univ. Press, 2016), 691; Sara Imari Walker, "The Origins of Life: A Problem for Physics, a Key Issues Review," *Reports on Progress in Physics* 80 (2017): 092601; Carl Woese, "The Universal Ancestor," *Proceedings of the National Academy of Sciences* 95, no. 12 (1998): 6854–6859.

Each of these disciplines has their own set of variables that they will focus on. Discovering life will take the combination of disciplines and their observables. Broadly speaking, we can detect life by seeing it directly (for example, a bug moving on a surface, or a plant growing or changing color), or by seeing signs it produces (for example, the presence of methane or oxygen in the atmosphere), or by seeing evidence of its past (such as fossil evidence or structural changes in topography). On Earth, the oldest fossil life has caused a structural topographic change with stromatolites—with one such fossil remnant created by cyanobacteria as large as six meters. Size matters: if it is microscopic evidence, it will take a microscope to observe it; if it is macroscopic evidence but still small, perhaps high-resolution observations can find it, but that will be hard. Something large, such as a forest or a herd of giraffes, will be much easier to observe—and much more unlikely. Biosignatures and surface features from living processes or ones that we associate with life are examples of indirect detections of life. For instance, we may obtain a multipixel image of an exoplanet at, say, 10 km resolution, and we may see unambiguously a liquid water ocean. It would not be proof of life, but it would be a strong suggestion of its likelihood (probably resulting in not just tabloid headlines, but even science journal headlines). If the multipixel image were of a grove of trees, or a large bed of stromatolites, that would be a direct detection of life. The ETI enthusiasts suggest we also look for technosignatures. Earlier in this book, I detailed why the existence of ETI is so unlikely—but unlikely doesn't mean it is impossible. Seeing smoke in a high-resolution image of an exoplanet would probably lead to the conclusion "where there is smoke, there is fire." That might indicate ETI burning something or just a forest fire, but in either case—life. A deeper quandary is raised by the questions discussed in chapter 3 about the future evolution of human life and of intelligence. What if the future of intelligence is largely machine-based? Might their technosignatures be uniquely noticeable? If we really were able to explore, up close (with multipixel imaging and other sensors, bringing that data to Earth), many potentially habitable planets—say dozens—we would have larger statistical samples and a wider range of conditions in which to search for life. Our conclusions about how life evolves and the likelihood of ETI therefore would be stronger. For this reason, comparative astrobiology will provide the data to test theories about the future of human evolution and intelligence (human, artificial, organic, inorganic) and perhaps give us insights as to how to deal with the existential threats now facing us.

Current exoplanet astrobiology science is focused on biosignatures in the atmosphere because that it is all that is possible to study within a one-pixel observation. These biosignatures are chemical signatures produced by molecules strongly associated with life—that is, with biological processes. In our own solar system where spacecraft currently reach the planets, the search for biosignatures also includes looking for surface features on the planet that result from biological processes or the movement of life. Implicitly, with biosignatures, I am talking about life more or less as we know it; I put aside for later discussion the question of life as we don't know it.

Biosignatures in Earth's atmosphere have been observed (of course) and, more than that, changes in some of them have been observed by remote sensing satellites. Some of the gas molecules identified as biosignatures are oxygen, ozone, methane, and nitrous oxide. Molecules are identified by analyzing spectral data—that is, emissions at different wavelengths. The association of a wavelength with the emission of a particular molecule or chemical combination is something learned from quantum physics. There is now a huge library of known spectral features of molecules. The spectrum of a planet is observed when light passes through its atmosphere and is absorbed by the molecule or retransmitted as an emission line, or both. One very important molecule connected with life (on Earth) is oxygen, and its detection is a huge goal for astrobiologists looking for biosignatures. We know how important it is on Earth to life processes. However, we also know life on Earth existed for a couple billion years before there was any appreciable oxygen, and we know oxygen is a product of other planetary processes (geological and chemical) without life. Thus, oxygen is a hopeful biosignature, but not a conclusive one—at least not by itself. In chapter 5, I cited a number of new telescopes being developed for space observations with greater and greater capabilities to observe exoplanets— still with a less-than-one-pixel limitation, but with increasing spectral sensitivity. I mentioned other gases whose presence is often associated with life—methane, for example. It is important however to realize that these gases, while possibly associated with life, may well be present due to nonbiologic processes on the exoplanet. For example, methane, while mostly produced from organic processes related to life, can come from natural gas pockets deep in Earth's mantle.

A review paper published in the journal *Astrobiology* stated this warning clearly, ". . . *all* hypothetical biosignatures could have false positives (nonbiological origins), and ground truth verification of the biogenicity of any remotely detected signature will be unachievable for the determinable future."[2] This underscores our conclusion that physically getting to exoplanets is a bridge too far, but it neglects our proposed use of the SGL as Nature's telescope to bring them closer up for observation.

The types of exoplanet biosignatures we can identify remotely are gas emission or absorption by life-associated molecules in the atmosphere, surface features (such as vegetation) that emit life-associated molecules as result of reflected light from the exoplanet's parent star, and observables that change with time and can be associated with life—for example, seasonal albedo changes or varying gas concentrations.

Biosignature data is then correlated with the concept of the habitable zone (discussed in chapter 5). However, that is a very superficial concept, based only on the temperature of the parent star and the location of the exoplanet in that star's system—that is, on the likely global temperature of the exoplanet's surface. Actually, the key parameter defining the habitable zone is the supposed ability of the planet in the zone to maintain liquid water on its surface. I have already noted how poorly that concept applies to our own system, where astrobiology studies rarely include Venus (also in the habitable zone) and emphatically include Europa, Enceladus, and Titan (well outside the habitable zone). The likely importance of liquid water makes direct observation of it a highly sought goal in the search for extraterrestrial life. If we could see the glint of reflected light from an exoplanet ocean—that would be a wow.[3] Any such claim from the less-than-one-pixel image we are currently limited to would be highly uncertain. Such an observation should be achievable with SGL imaging.

I noted above that Copernicus's discovery of a Sun-centered solar system didn't initially affect things much in the sixteenth century, but in the centuries that followed, it led to a whole new era in mathematics driven by the calculation of celestial orbits, which in turn contributed to developments in physics. In the one hundred years following Copernicus's publication, a lot more data and analysis came into play—Tycho Brahe's observations of the planets, Kepler's formulation of the laws of orbital mechanics, and, of course, Galileo's telescope observations of the Jovian satellites and Newton's law of gravity. Ultimately, it led to space flight, from celestial mechanics to rockets to spacecraft to scientific discoveries on other worlds. This look back at the impact of and follow-ups to Copernicus's heliocentric "discovery" may help us look ahead to the impact of and follow-ups to discoveries of extraterrestrial life. The discoveries about life on other worlds should have about the same kind of effect—after the first few hyped media events, the development of a new scientific field will result in a new understanding about life—on Earth and in the universe. Comparative astrobiology will become a new field that looks at processes, biosignatures, chemical reactions, and—if we succeed in using the solar gravity lens—surface features related to life on many different worlds. Multiple disciplines will be involved, as listed in table 3 (pp. 98–100). The nondetections will also contribute to our understanding of the processes involved, and, literally, multiple planet observations will serve as astrobiology laboratories for testing different hypotheses and putting together theories about the origin and evolution of life. By the beginning of the next century, we should have in-situ data from several astrobiological targets of interest in our own solar system (at least Mars, Europa, Titan, and Venus), as well as observations of life or signs of life on several exoplanets. That is, if we can conduct "real" interstellar exploration on multiple exoplanet SGL focus lines as described in the previous chapter. And, if we are really doing this with advanced twenty-first-century VR and AI tools, we will be looking at all these worlds and comparing their life processes (or, in the absence of life, their prebiotic chemistries) in Earth-based facilities with thousands of scientists and an interested and engaged public.[4]

This vision of exploring other worlds and interacting with other life throughout our solar system and the Galaxy is why I am so optimistic about and excited by the new age of exploration that we are opening up, forgetting interstellar voyage fantasies and wishing to contact extraterrestrials and instead using new engineering tools and new understandings of science. Reality will win out over fiction.

APPENDIX A

INTERSTELLAR MESSAGING

MESSAGES COME IN different forms—there is voice mail, email, text, radio and television advertising, and the many forms of social media. The electromagnetic ones (as opposed to print) are transmitted with zero mass. There are also postal mail, delivery services, billboards, murals, and the proverbial message in a bottle. These have mass—relying on hardware more than software. The same applies in space, specifically in interstellar space: we could have radio and optical electromagnetic-wave messages, and we could have space probes carrying plaques, discs with recordings, and microetched lithographs.

Photons travel at the speed of light, which is, by definition, one light-year per year, or approximately 270,000 AU per year (or 7.2 AU per hour, a unit useful for computing communication times in the solar system). Light speed applies not just to light but to all electromagnetic radiation in a vacuum, specifically radio waves, which are also used for communication. Spacecraft carrying messages travel much more slowly—Voyager is the fastest so far, carrying the Voyager Golden Record into interstellar space at about 3.2 AU per year. In chapter 7, means of achieving speeds greater than 20 AU per year—perhaps even as high as 50 AU per year—were discussed.

In chapter 3, it was concluded that the "eerie silence" (as scientist Paul Davies describes it[1]) is not surprising because ETI is unlikely—that is, the evolution of intelligence with communications technology is a highly unlikely happenstance. Then, in chapter 6, I showed that interstellar travel is practically impossible—distances are too large, energy requirements too great, and, even if technology were stretched as far as we can think, the results far too limited to be useful. Why then even think about interstellar messaging? The most important thing to understand about interstellar messaging *is that it has little to do with the receiver and much to do with the sender*. This is true for both those who accept this book's conclusion— that we are likely alone in the universe—and those who remain convinced there is intelligent life "out there" with which we can imagine communicating. Communication is a two-way process while messaging goes one way, but both implicitly assume there is a sender and a receiver. SETI, discussed in chapter 2, assumes an intelligent (technological) sender on some other world either directing a message toward us or broadcasting their message widely to the universe. A third possibility (more prevalent on Earth) is that a message might not be purposeful, but accidental—just leakage from their active transmitters and other technological activity on their planet. On Earth, we have been looking for interstellar messages for more than a half century. We also are sending out a lot of signals that broadcast our existence at many wavelengths, including light and radio/television. SETI itself is not too controversial (except politically when its funding is debated), but purposefully broadcasting signals from Earth is akin to shouting in a dark, unknown, and dangerous forest (it is called the dark forest theory).[2] Some scientists push to ban efforts to purposely broadcast interstellar messages, but most consider any proposed ban as unenforceable or useless given the magnitude of our already active broadcasting of radio, television, radar, and light signals. If the universe really has a lot of other observers, it is hard to think of us remaining hidden.

The idea of sending a message into space goes back to before even the advent of communications technology. Mathematician Carl Friedrich Gauss proposed in 1820 to create a right triangle of cleared trees in the Siberian forest that could possibly be seen by beings on another planet. In the 1840s, astronomer Joseph Johann von Littrow suggested doing something similar by lighting huge fires in trenches in the Sahara. In the early days of electricity, telegraphs, and radio, inventors Nikola Tesla and Guglielmo Marconi suggested two-way communications with beings on other planets was possible, and even claimed to receive some signals from Mars. However, it was the invention and development of the radio telescope that led to the scientific study of extraterrestrial communication. Almost all of that has been within the discipline of SETI (listening for radio signals or watching for optical signals). The idea of messaging took off in the 1970s with astronomer Frank Drake, then professor of astronomy at Cornell University. Cornell operated Arecibo, the largest radio telescope in the world, in Puerto Rico, and the technical intent of Drake's project was to demonstrate the power of its radar transmitter. Drake led a group, which included Carl Sagan, that devised the contents of the message—a binary code of ones and zeros that represented significant information about physics, chemistry, and biology to identify us as an intelligent species, and even stated our location in the universe. If that seems dangerous, it also highlights the inability to ban signals—Drake's team just did it. The transmission did indeed prove the power of radio astronomy, and it led to the whole field of SETI. But, to anyone's knowledge, no one ever received it.[3] The SETI scientific community has largely abandoned the notion of purposely sending messages for the reasons stated above, but amateur groups with access to large telescopes continue to try. One such group was Team Encounter, who rented time on the large Ukrainian radio telescope in Yevpatoria and sent out a number of signals in 1999. These signals are also on their way out into the universe, purportedly with both scientific and public content. However, this scheme got no serious scientific backing, and public involvement was minimal. Team Encounter later tried to raise money for a space mission to head out of the solar system with people's DNA, but it never got off the ground. Yevpatoria was the main radio telescope for the Soviet Union's planetary missions, and earlier in 1962 a group of Soviet scientists used it to broadcast a message to Venus in advance of space missions going there.[4] Another message was sent by NASA (!) that consisted of the Beatles song "Across the Universe." It was sent in 2009 from the Madrid 70-Meter Antenna, which was part of the NASA Deep Space Network. It is now winging its way across the Galaxy.

The first space probes able to head out of the solar system toward interstellar space were the Pioneers (10 and 11), launched in 1972 and 1973 to fly by Jupiter and Saturn and use their gravity to boost their speed to exit the solar system.[5] Eric Burgess, a British journalist, author, and space aficionado (a charter member of the British Interplanetary Society) visited JPL in Pasadena, California, in 1971 (when Mariner 9 went into orbit around Mars) and suggested to Carl Sagan, then working with the Mariner science team, that the Pioneer spacecraft should carry a message along their interstellar journey. Carl had led an international conference about SETI and was working with Frank Drake at Cornell to formulate the Arecibo message. He convinced NASA to include a plaque on the Pioneer 10 spacecraft, and when it launched in 1972 it became the first interstellar message of the Space Age (the Arecibo message was transmitted in 1974).[6] The plaque (on both Pioneer 10 and 11), developed by Sagan and Drake, was a pictorial representation of who we (Earthlings) are and where we come from. It very much was in the spirit of a message in a bottle tossed into the ocean of space by a young species contemplating its interstellar journey. A message from a young species who wondered about what is beyond the solar system, without expecting an answer or an encounter.

The Pioneer plaques were followed by the Voyager record. Pioneer itself, a rather low-cost spinning spacecraft with a limited payload and imaging capability, was followed by Voyager, a much larger three-axis-stabilized spacecraft with much more imaging capability. Voyager was a downsized version of an even more ambitious plan called Grand Tour, designed to have a fully autonomous spacecraft fly by the four large outer planets, Jupiter, Saturn, Uranus, and Neptune. The cost estimate was too big for the NASA budget, so Grand Tour was cancelled, and a group of JPL engineers worked fast and furiously for a couple of weeks to come up with a lower-cost alternative to fit into the budget. The result was called Mariner Jupiter-Saturn, a two-planet, not-so-grand tour. However, they designed it so well that they kept open options for survival after Saturn, and ultimately the two Voyager spacecraft visited all four of those large outer planets, accomplishing the grand tour after all. Similarly, the Voyager record was an expansion of the simpler Arecibo message and Pioneer plaque—Carl Sagan brought a much broader group together to develop the record to have a lot more information. Music, art, speech recordings, and other representations of Earth (from all corners) were included on the Golden Record. Voyager has no specific interstellar destination, but the idea (as with the plaque and Arecibo message) is that some species might receive it. However, as noted earlier, interstellar messages are not about the putative recipient; they are about the sender. The president of the United States at the time of Voyager's launch, Jimmy Carter, said, "This is a present from a small, distant world, a token of our sounds, our science, our images, our music, our thoughts and our feelings. We are attempting to survive our time so we may live into yours," clearly making the message about "us," not "them."[7] The impact of the Voyager record in popular culture illustrates the potency of the interstellar message, even without an expected recipient.

The Planetary Society picked up on this idea in the 1990s and developed what became the first privately funded payloads to another world. First, they sent a nanolithograph on the Mars Pathfinder mission (in 1996) and then a DVD collection of science fiction stories, art, and essays on Phoenix in 2007 (originally intended for a Russian Mars 1996 mission) to the surface of Mars. The Pathfinder Microelectronics and Photonics Experiment (MAPEx) with a radiation recorder that measures the radiation environment and a list of Planetary Society member names sits still at the Pathfinder landing site (now named the Carl Sagan Memorial Station), waiting for a human to get there and pick it up. The Visions of Mars Library also sits on Mars, at the site of the Phoenix spacecraft in the north polar region, similarly waiting for humans on Mars to find it. Both projects were representations of Earthly expression far more than they were expected to be really found any time soon. That is the essence of interstellar (or in these cases planetary) messaging.

What about future missions? Chapter 8 discussed real interstellar exploration and proposed a series of ever-faster and more capable missions that could explore exoplanets by the end of this century. It is reasonable to expect that they discover life—not life to communicate with, but life to learn from. These missions should have public engagement and participation. One of the missions in that proposed series was a rendezvous and close examination of an interstellar object. We might think of ISOs as messages from the stars—not alien spacecraft, (as in the eccentric idea discussed in chapter 6), but natural phenomena from other star systems that we might examine. The solar sail–smallsat can catch such objects, even if they are discovered only months before they enter the inner solar system. If we send a probe onto the surface of an ISO and attach it there, it will ride with it out of the solar system. Such a mission should have a message, ideally one with a radio or optical beacon that allows it to be tracked as it flies into interstellar space. Of course, the probe would also land science instruments to measure the texture and composition of the surface and to determine its mass, shape, and density. That information would be transmitted to Earth, and we would then be able to determine the similarities and differences of this ISO to objects in our own solar system.

A variation on the theme of using a comet for messaging is being developed by a team led by Greg Pass at Cornell University and Dan Goods at JPL: the Altamira Comet.[8] Their idea is to collect more than one billion portraits from Earth's population and process them by electron beam lithography into 3D gold drawings and place them into a ball of water about the size of a baseball. The water is then frozen and, when taken to space and released far out from the Sun, it becomes an artificial comet in which each individual drawing is a speck of dust, like the dust from a comet's nucleus. This would be a worldwide public involvement project sending a sign from humanity into the future. The Altamira caves in Spain contain such representations in the forms of handprints from prehistoric people—their sign of existence for the future.

Another type of probe is a Bracewell probe, a concept for an autonomous interstellar probe designed to communicate with alien species. It was proposed by Prof. Ronald N. Bracewell of Stanford University in 1960 as a faster means of communication than radio transmission from Earth. The basic idea is to construct the probe's computer with everything we want to communicate about ourselves.[9] With advanced AI, this computer could interact with whatever it finds on its interstellar voyage—an alien species, or even just a world with primitive life. In either case, all the consequences of the interaction fall on the alien world—there is presumably no communication with whatever remains here on Earth after the many centuries of flight to wherever the Bracewell probe has gone. If that probe is also a von Neumann probe, that is, a robotic probe that can reproduce itself indefinitely, then this concept morphs into one in which robotic emissaries (or, should we say, descendants) embark for thousands or tens of thousands of years and interact with alien worlds. This whole notion is counter to the ideas of this book, as it presupposes a universe teeming with intelligent life capable of communication and interaction with our emissary. It's a great science fiction theme, but there is no basis for it in science. Personally, I would wait a century or two before deciding about our robotic descendants to learn a little more first.

The subject of interstellar messaging must be approached with care and critical thinking. It is credible to send a collection of art and literature that represent Earth as an interstellar communication, but not to light large fires on Earth to be detected by alien species. The former has an educational and cultural connection to our interest in exploring the universe, while the latter has no scientific basis and results in bad environmental fallout without any redeeming benefit. As a cofounder and longtime executive director of The Planetary Society, I strongly believe in public engagement in space exploration. Exploration, not dilettantism. The engagement should support the science of exploration and be credible (technically and economically feasible) in its own right. Most importantly, it should be honest. In that spirit, I hope there will be more messages on future missions, engagement that contributes a bit to understanding more about the universe and a lot to understanding ourselves.

APPENDIX B

MAYR VERSUS SAGAN, A DEBATE ON EXTRATERRESTRIAL INTELLIGENCE

This is a reprint from a 1996 issue of the Planetary Report. *It is included here because the two brilliant and distinguished scientists illustrate well the differing opinions of astronomers and biologists about the likelihood of extraterrestrial intelligence.*

THE SEARCH FOR EXTRATERRESTRIAL INTELLIGENCE:
SCIENTIFIC QUEST OR HOPEFUL FOLLY?

A DEBATE BETWEEN ERNST MAYR AND CARL SAGAN

S INCE HUMANS FIRST looked up, they have seen in the skies the phantoms of their wondering minds. If there is one thread that links the ancient Greek philosophers to modern space scientists, it is the uncertainty about the plurality of worlds. Vast and ancient beyond ordinary human understanding, the universe leaves us pondering the ultimate significance, if any, of our tiny but exquisite life-bearing blue planet.

With the development of technology and our present understanding of the laws of nature, the human species is now in a position where the possibility of extraterrestrial civilizations can be verified by experiment. But because we have yet to find a single piece of concrete evidence of alien intelligence, a philosophical battle has arisen between those who might be called contact optimists—who generally embrace SETI—and the proponents of the uniqueness hypothesis, which suggests that Earth is the only technical civilization in our Galaxy.

In these pages, we present both sides of this philosophical and scientific battle. Which view is more palatable to you? Read on and decide for yourself.

—Guillermo A. Lemarchand

THE IMPROBABILITY OF SUCCESS
BY ERNST MAYR

What is the chance of success in the Search for Extraterrestrial Intelligence?

The answer to this question depends on a series of probabilities. I have attempted to make a detailed analysis of this problem in a German publication (Mayr, 1992) and shall attempt here to present in English the essential findings of this investigation. My methodology consists in asking a series of questions that narrow down the probability of success.

How probable is it that life exists somewhere else in the universe?

Even most skeptics of the SETI project will answer this question optimistically. Molecules that are necessary for the origin of life, such as amino acids and nucleic acids, have been identified in cosmic dust, together with other macromolecules, and so it would seem quite conceivable that life could originate elsewhere in the universe.

Some of the modern scenarios of the origin of life start out with even simpler molecules—a beginning that makes an independent origin of life even more probable. Such an independent origin of life, however, would presumably result in living entities that are drastically different from life on Earth.

Where can one expect to find such life?

Obviously, only on planets. Even though we have up to now secure knowledge only of the nine planets of our solar system, there is no reason to doubt that in all galaxies there must be millions if not billions of planets. The exact figure, for instance, for our own Galaxy can only be guessed.

How many of these planets would have been suitable for the origin of life?

There are evidently rather narrow constraints for the possibility of the origin and maintenance of life on a planet. There has to be a favorable average temperature; the seasonal variation should not be too extreme; the planet must have a suitable distance from its sun; it must have the appropriate mass so that its gravity can hold an atmosphere; this atmosphere must have the right chemical composition to support early life; it must have the necessary consistency to protect the new life against ultraviolet and other harmful radiations; and there must be water on such a planet. In other words, all environmental conditions must be suitable for the origin and maintenance of life.

One of the nine planets of our solar system had the right kind of mixture of these factors. This, surely, was a matter of chance. What fraction of planets in other solar systems will have an equally suitable combination of environmental factors? Would it be one in 10, or one in 100, or one in 1,000,000? Which figure you choose depends on your optimism. It is always difficult to extrapolate from a single instance. This figure, however, is of some importance when you are dealing with the limited number of planets that can be reached by any of the SETI projects.

What percentage of planets on which life has originated will produce intelligent life?

Physicists, on the whole, will give a different answer to this question than biologists. Physicists still tend to think more deterministically than biologists. They tend to say that if life has originated somewhere, it will also develop intelligence in due time. The biologist, on the other hand, is impressed by the improbability of such a development.

Life originated on Earth about 3.8 billion years ago, but high intelligence did not develop until about half a million years ago. If Earth had been temporarily cooled down or heated up too much during these 3.8 billion years, intelligence would have never originated.

When answering this question, one must be aware of the fact that evolution never moves in a straight line toward an objective ("intelligence"), as happens during a chemical process or as a result of a law of physics. Evolutionary pathways are highly complex and resemble more a tree with all of its branches and twigs.

After the origin of life—that is, 3.8 billion years ago—life on Earth consisted for 2 billion years only of simple prokaryotes, cells without an organized nucleus. These bacteria and their relatives developed surely 50 to 100 different (some perhaps very different) lineages, but, in this enormously long time, none of them led to intelligence. Owing to an astonishing, unique event that is even today only partially explained, about 1,800 million years ago the first eukaryote originated, a creature with a well-organized nucleus and the other characteristics of "higher" organisms. From the rich world of the protists (consisting of only a single cell), there eventually originated three groups of multicellular organisms: fungi, plants, and animals. But none of the millions of species of fungi and plants was able to produce intelligence.

The animals (Metazoa) branched out in the Precambrian and Cambrian time periods to about 60 to 80 lineages (phyla). Only a single one of them, that of the chordates, led eventually to genuine intelligence. The chordates are an old and well diversified group, but only one of its numerous lineages, that of the vertebrates, eventually produced intelligence. Among the vertebrates, a whole series of groups evolved—types of fishes, amphibians, reptiles, birds, and mammals. Again, only a single lineage, that of the mammals, led to high intelligence. The mammals had a long evolutionary history which began in the Triassic Period, more than 200 million years ago, but only in the latter part of the Tertiary Period—that is, some 15 to 20 million years ago—did higher intelligence originate in one of the circa 24 orders of mammals.

The elaboration of the brain of the hominids began less than 3 million years ago, and that of the cortex of *Homo sapiens* occurred only about 300,000 years ago. Nothing demonstrates the improbability of the origin of high intelligence better than the millions of phyletic lineages that failed to achieve it.

How many species have existed since the origin of life?

This figure is as much a matter of speculation as the number of planets in our Galaxy. But if there are 30 million living species, and if the average life expectancy of a species is about 100,000 years, then one can postulate that there have been billions, perhaps as many as 50 billion species since the origin of life. Only one of these achieved the kind of intelligence needed to establish a civilization.

To provide exact figures is difficult because the range of variation both in the origination of species and in their life expectancy is so enormous. The widespread, populous species of long geological duration (millions of years), usually encountered by the paleontologist, are probably exceptional rather than typical.

Why is high intelligence so rare?

Adaptations that are favored by selection, such as eyes or bioluminescence, originate in evolution scores of times independently. High intelligence has originated only once, in human beings. I can think of only two possible reasons for this rarity. One is that high intelligence is not at all favored by natural selection, contrary to what we would expect. In fact, all the other kinds of living organisms, millions of species, get along fine without high intelligence.

The other possible reason for the rarity of intelligence is that it is extraordinarily difficult to acquire. Some grade of intelligence is found only among warm-blooded animals (birds and mammals), not surprisingly so because brains have extremely high energy requirements. But it is still a very big step from "some intelligence" to "high intelligence."

The hominid lineage separated from the chimpanzee lineage about 5 million years ago, but the big brain of modern man was acquired less than 300,000 years ago. As one scientist has suggested (Stanley, 1992), it required complete emancipation from arboreal life to make the arms of the mothers available to carry the helpless babies during the final stages of brain growth. Thus, a large brain, permitting high intelligence, developed in less than the last 6 percent of the life on the hominid line. It seems that it requires a complex combination of rare, favorable circumstances to produce high intelligence (Mayr, 1994).

How much intelligence is necessary to produce a civilization?

As stated, rudiments of intelligence are found already among birds (ravens, parrots) and among nonhominid mammals (carnivores, porpoises, monkeys, apes, and so forth), but none of these instances of intelligence has been sufficient to found a civilization.

Is every civilization able to send signals into space and to receive them?

The answer quite clearly is no. In the last 10,000 years, there have been at least 20 civilizations on Earth, from the Indus, the Sumerian, and other Near Eastern civilizations, to Egypt, Greece, and the whole series of European civilizations, to the Mayas, Aztecs, and Incas, and to the various Chinese and Indian civilizations. Only one of these reached a level of technology that has enabled it to send signals into space and to receive them.

Would the sense organs of extraterrestrial beings be adapted to receive our electronic signals?

This is by no means certain. Even on Earth, many groups of animals are specialized for olfactory or other chemical stimuli and would not react to electronic signals. Neither plants nor fungi are able to receive electronic signals. Even if there were higher organisms on some planet, it would be rather improbable that they would have developed the same sense organs that we have.

How long is a civilization able to receive signals?

All civilizations have only a short duration. I will try to emphasize the importance of this point by telling a little fable.

Let us assume that there were really intelligent beings on another planet in our Galaxy. A billion years ago, their astronomers discovered Earth and reached the conclusion that this planet might have the proper conditions to produce intelligence. To test this, they sent signals to Earth for a billion years without ever getting an answer. Finally, in the year 1800 (of our calendar) they decided they would send signals only for another 100 years. By the year 1900, no answer had been received, so they concluded that surely there was no intelligent life on Earth.

This shows that even if there were thousands of civilizations in the universe, the probability of a successful communication would be extremely slight because of the short duration of the "open window."

One must not forget that the range of SETI systems is very limited, reaching only part of our Galaxy. The fact that there are a near infinite number of additional galaxies in the universe is irrelevant as far as SETI projects are concerned.

SETI Success: An Improbability of Astronomic Dimensions

What conclusions must we draw from these considerations? No less than six of the eight conditions to be met for SETI success are improbable. When one multiplies these six improbabilities with each other, one reaches an improbability of astronomic dimensions.

Why are there nevertheless still proponents of SETI? When one looks at their qualifications, one finds that they are almost exclusively astronomers, physicists, and engineers. They are simply unaware of the fact that the success of any SETI effort is not a matter of physical laws and engineering capabilities but essentially a matter of biological and sociological factors. These, quite obviously, have been entirely left out of the calculations of the possible success of any SETI project.

REFERENCES

Ernst Mayr, "Lohnt sich die Suche nach extraterrestrischer Intelligenz" [Is It Worthwhile to Search for Extraterrestrial Intelligence?]. *Naturwissenschaftliche Rundschau* 45, no. 7 (1992), 264–266.

Ernst Mayr, "Does It Pay to Acquire High Intelligence?" *Perspectives in Biology and Medicine* (1994), 150–154.

Steven M. Stanley, "An Ecological Theory for the Origin of *Homo*." *Paleobiology* 18 (1992), 237–257.

THE ABUNDANCE OF LIFE-BEARING PLANETS
BY CARL SAGAN

We live in an age of remarkable exploration and discovery. Fully half of the nearby Sun-like stars have circumstellar disks of gas and dust like the solar nebula out of which our planets formed 4.6 billion years ago. By a most unexpected technique—radio timing residuals—we have discovered two Earth-mass planets around the pulsar B1257+12. Apparent Jovian planets have been detected around the stars 51 Pegasi, 70 Virginis, and 47 Ursae Majoris. A range of new Earth-based and spaceborne techniques— including astrometry, spectrophotometry, radial velocity measurements, adaptive optics and interferometry—all seem to be on the verge of being able to detect Jovian-type planets, if they exist, around the nearest stars. At least one proposal (The FRESIP [Frequency of Earth-Sized Inner Planets] Project, a spaceborne spectrophotometric system) holds the promise of detecting terrestrial planets more readily than Jovian ones. If there is not a sudden cutoff in support, we are likely entering a golden age in the study of the planets of other stars in the Milky Way galaxy.

Once you have found another planet of Earthlike mass, however, it of course does not follow that it is an Earthlike world. Consider Venus. But there are means by which, even from the vantage point of Earth, we can investigate this question. We can look for the spectral signature of enough water to be consistent with oceans. We can look for oxygen and ozone in the planet's atmosphere. We can seek molecules like methane, in such wild thermodynamic disequilibrium with the oxygen that it can only be produced by life. (In fact, all of these tests for life were successfully performed by the Galileo spacecraft in its close approaches to Earth in 1990 and 1992 as it wended its way to Jupiter [Sagan et al., 1993].)

The best current estimates of the number and spacing of Earth-mass planets in newly forming planetary systems (as George Wetherill reported at the first international conference on circumstellar habitable zones [Doyle, 1996]) combined with the best current estimates of the long-term stability of oceans on a variety of planets (as James Kasting reported at that same meeting [Doyle, 1996]) suggest one to two blue worlds around every Sun-like star. Stars much more massive than the Sun are comparatively rare and age quickly. Stars comparatively less massive than the Sun are expected to have Earthlike planets, but the planets that are warm enough for life are probably tidally locked so that one side always faces the local sun. However, winds may redistribute heat from one hemisphere to another on such worlds, and there has been very little work on their potential habitability.

Nevertheless, the bulk of the current evidence suggests a vast number of planets distributed through the Milky Way with abundant liquid water stable over lifetimes of billions of years. Some will be suitable for life—our kind of carbon and water life—for billions of years less than Earth, some for billions of years more. And, of course, the Milky Way is one of an enormous number, perhaps a hundred billion, other galaxies.

Need Intelligence Evolve on an Inhabited World?

We know from lunar cratering statistics, calibrated by returned *Apollo* samples, that Earth was under hellish bombardment by small and large worlds from space until around 4 billion years ago. This pummeling was sufficiently severe to drive entire atmospheres and oceans into space. Earlier, the entire crust of Earth was a magma ocean. Clearly, this was no breeding ground for life.

Yet, shortly thereafter—Mayr adopts the number 3.8 billion years ago—some early organisms arose (according to the fossil evidence). Presumably the origin of life had to have occurred some time before that. As soon as conditions were favorable, life began amazingly fast on our planet. I have used this fact (Sagan, 1974) to argue that the origin of life must be a highly probable circumstance; as soon as conditions permit, up it pops!

Now, I recognize that this is at best a plausibility argument and little more than an extrapolation from a single example. But we are data constrained; it's the best we can do.

Does a similar analysis apply to the evolution of intelligence? Here you have a planet burgeoning with life, profoundly changing the physical environment, generating an oxygen atmosphere 2 billion years ago, going through the elegant diversification that Mayr briefly summarized—and not for almost 4 billion years does anything remotely resembling a technical civilization emerge.

In the early days of such debates (for example, G. G. Simpson's "The Non-prevalence of Humanoids"), writers argued that an enormous number of individually unlikely steps were required to produce something very like a human being, a "humanoid"; that the chances of such a precise repetition occurring on another planet were nil; and therefore, that the chance of extraterrestrial intelligence was nil. But clearly when we're talking about extraterrestrial intelligence, we are not talking—despite *Star Trek*—of humans or humanoids. We are talking about the functional equivalent of humans—say, any creatures able to build and operate radio telescopes. They may live on the land or in the sea or in the air. They may have unimaginable chemistries, shapes, sizes, colors, appendages, and opinions. We are not requiring that they follow the route that led to the evolution of humans. There may be many different evolutionary pathways, each unlikely, but the sum of the number of pathways to intelligence may nevertheless be quite substantial.

In Mayr's current presentation, there is still an echo of "the non-prevalence of humanoids." But the basic argument is, I think, acceptable to all of us. Evolution is opportunistic and not foresighted. It does not "plan" to develop intelligent life a few billion years into the future. It responds to short-term contingencies. And yet, other things being equal, it is better to be smart than to be stupid, and an overall trend toward intelligence can be perceived in the fossil record. On some worlds, the selection pressure for intelligence may be higher, on others, lower.

If we consider the statistics of one, our own case—and take a typical time from the origin of a planetary system to the development of a technical civilization to be 4.6 billion years—what follows? We would not expect civilizations on different worlds to evolve in lockstep. Some would reach technical intelligence more quickly, some more slowly, and—doubtless—some never. But the Milky Way is filled with second- and third-generation stars (that is, those with heavy elements) as old as 10 billion years.

So let's imagine two curves: The first is the probable timescale to the evolution of technical intelligence. It starts out very low; by a few billion years it may have a noticeable value; by 5 billion years, it's something like 50 percent; by 10 billion years, maybe it's approaching 100 percent. The second curve is the ages of Sun-like stars, some of which are very young—they're being born right now—some of which are as old as the Sun, some of which are 10 billion years old. If we convolve these two curves, we find there's a chance of technical civilizations on planets of stars of many different ages—not much in the very young ones, more and more for the older ones. The most likely case is that we will hear from a civilization considerably more advanced than ours. For each of those technical civilizations, there will have been tens of billions or more other species. The number of unlikely events that had to be concatenated to evolve a technical species is enormous, and perhaps there are members of each of those species who pride themselves on being uniquely intelligent in all the universe.

Need Civilizations Develop the Technology for SETI?

It is perfectly possible to imagine civilizations of poets or (perhaps) Bronze Age warriors who never stumble on James Clerk Maxwell's equations and radio receivers. But they are removed by natural selection. The Earth is surrounded by a population of asteroids and comets, such that occasionally the planet is struck by one large enough to do substantial damage.

The most famous is the K-T event (the massive near-Earth object impact that occurred at the end of the Cretaceous Period and start of the Tertiary) of 65 million years ago that extinguished the dinosaurs and most other species of life on Earth. But the chance is something like one in 2,000 that a civilization-destroying impact will occur in the next century.

It is already clear that we need elaborate means for detecting and tracking near-Earth objects and the means for their interception and destruction. If we fail to do so, we will simply be destroyed. The Indus Valley, Sumerian, Egyptian, Greek, and other civilizations did not have to face this crisis because they did not live long enough. Any long-lived civilization, terrestrial or extraterrestrial, must come to grips with this hazard. Other solar systems will have greater or lesser asteroidal and cometary fluxes, but in almost all cases the dangers should be substantial.

Radiotelemetry, radar monitoring of asteroids, and the entire concept of the electromagnetic spectrum are part and parcel of any early technology needed to deal with such a threat. Thus, any long-lived civilization will be forced by natural selection to develop the technology of SETI. (And there is no need to have sense organs that "see" in the radio region. Physics is enough.)

Since perturbation and collision in the asteroid and comet belts are perpetual, the asteroid and comet threat is likewise perpetual, and there is no time when the technology can be retired. Also, SETI itself is a small fraction of the cost of dealing with the asteroid and comet threat.

(Incidentally, it is by no means true that SETI is "very limited, reaching only part of our Galaxy." If there were sufficiently powerful transmitters, we could use SETI to explore distant galaxies; because the most likely transmitters are ancient, we can expect them to be powerful. This is one of the strategies of the Megachannel Extraterrestrial Assay [META].)

Is SETI a Fantasy of Physical Scientists?

Mayr has repeatedly suggested that proponents of SETI are almost exclusively physical scientists and that biologists know better. Since the relevant technologies involve the physical sciences, it is reasonable that astronomers, physicists, and engineers play a leading role in SETI.

But in 1982, when I put together a petition published in *Science* urging the scientific respectability of SETI, I had no difficulty getting a range of distinguished biologists and biochemists to sign, including David Baltimore, Melvin Calvin, Francis Crick, Manfred Eigen, Thomas Eisner, Stephen Jay Gould, Matthew Meselson, Linus Pauling, David Raup, and Edward O. Wilson. In my early speculations on these matters, I was much encouraged by the strong support from my mentor in biology, H. J. Muller, a Nobel laureate in genetics.

The petition proposed that, instead of arguing the issue, we look: "We are unanimous in our conviction that the only significant test of the existence of extraterrestrial intelligence is an experimental one. No a priori arguments on this subject can be compelling or should be used as a substitute for an observational program."

REFERENCES

L. R. Doyle, ed., "Circumstellar Habitable Zones: Proceedings of the First International Conference," Travis House Publications, Menlo Park, California, 1996.

Carl Sagan, "The Origin of Life in a Cosmic Context." *Origins of Life* 5 (1974), 497–505.

Carl Sagan and Ann Druyan, *Shadows of Forgotten Ancestors: A Search for Who We Are* (New York: Random House, 1992).

Carl Sagan et al., "A Search for Life on Earth from the Galileo Spacecraft." *Nature* 365 (1993), 715–721.

Carl Sagan, *Pale Blue Dot: A Vision of the Human Future in Space* (New York: Random House, 1994).

I. S. Shklovskii and Carl Sagan, *Intelligent Life in the Universe* (San Francisco: Holden-Day, 1966).

G. G. Simpson, "The Non-prevalence of Humanoids." *Science* 143 (1964), 769–775.

ERNST MAYR RESPONSE

I fully appreciate that the nature of our subject permits only probabilistic estimates. There is no argument between Carl Sagan and myself as to the probability of life elsewhere in the universe and the existence of large numbers of planets in our and other nearby galaxies. The issue, as correctly emphasized by Sagan, is the probability of the evolution of high intelligence and an electronic civilization on an inhabited world.

Once we have life (and almost surely it will be very different from life on Earth), what is the probability of its developing a lineage with high intelligence? On Earth, among millions of lineages of organisms and perhaps 50 billion speciation events, only one led to high intelligence; this makes me believe in its utter improbability.

Sagan adopts the principle "it is better to be smart than to be stupid," but life on Earth refutes this claim. Among all the forms of life, neither the prokaryotes nor protists, fungi or plants have evolved smartness, as they should have if they were "better." In the 28-plus phyla of animals, intelligence evolved in only one (chordates) and doubtfully also in the cephalopods. And in the thousands of subdivisions of the chordates, high intelligence developed in only one, the primates, and even there only in one small subdivision. So much for the putative inevitability of the development of high intelligence because "it is better to be smart."

Sagan applies physicalist thinking to this problem. He constructs two linear curves, both based on strictly deterministic thinking. Such thinking is often quite legitimate for physical phenomena but is quite inappropriate for evolutionary events or social processes such as the origin of civilizations. The argument that extraterrestrials, if belonging to a long-lived civilization, will be forced by selection to develop an electronic know-how to meet the peril of asteroid impacts is totally unrealistic How would the survivors of earlier impacts be selected to develop the electronic know-how? Also, the case of Earth shows how impossible the origin of any civilization is unless high intelligence develops first. Earth furthermore shows that civilizations inevitably are short-lived.

It is only a matter of common sense that the existence of extraterrestrial intelligence cannot be established by a priori arguments. But this does not justify SETI projects, since it can be shown that the success of an observational program is so totally improbable that it can, for all practical purposes, be considered zero.

All in all, I do not have the impression that Sagan's rebuttal has weakened in any way the force of my arguments.

CARL SAGAN RESPONSE

The gist of Professor Mayr's argument is essentially to run through the various factors in the Drake equation (see Shklovskii and Sagan, 1966) and attach qualitative values to each. He and I agree that the probabilities concerning the abundance of planets and the origins of life are likely to be high. (I stress again that the latest results [Doyle, 1996] suggest one or even two Earth-like planets with abundant surface liquid water in each planetary system. The conclusion is of course highly tentative, but it encourages optimism.) Where Mayr and I disagree is in the later factors in the Drake equation, especially those concerning the likelihood of the evolution of intelligence and technical civilizations.

Mayr argues that prokaryotes and protista have not "evolved smartness." Despite the great respect in which I hold Professor Mayr, I must demur: Prokaryotes and protista are our ancestors. They *have* evolved smartness, along with most of the rest of the gorgeous diversity of life on Earth.

On the one hand, when he notes the small fraction of species that have technological intelligence, Mayr argues for the relevance of life on Earth to the problem of extraterrestrial intelligence. But on the other hand, he neglects the example of life on Earth when he ignores the fact that intelligence has arisen here when our planet has another 5 billion years more evolution ahead of it. If it were legitimate to extrapolate from the one example of planetary life we have before us, it would follow that

1. There are enormous numbers of Earth-like planets, each stocked with enormous numbers of species, and
2. In much less than the stellar evolutionary lifetime of each planetary system, at least one of those species will develop high intelligence and technology.

Alternatively, we could argue that it is improper to extrapolate from a single example. But then Mayr's one-in-50 billion argument collapses. It seems to me he cannot have it both ways.

On the evolution of technology, I note that chimpanzees and bonobos have culture and technology. They not only use tools but also purposely manufacture them for future use (cf. Sagan and Druyan, 1992). In fact, the bonobo Kanzi has discovered how to manufacture stone tools.

It is true, as Mayr notes, that of the major human civilizations, only one has developed radio technology. But this says almost nothing about the probability of a human civilization developing such technology. That civilization with radio telescopes has also been at the forefront of weapons technology. If, for example, Western European civilization had not utterly destroyed Aztec civilization, would the Aztecs eventually—in centuries or millennia—have developed radio telescopes? They already had a superior astronomical calendar to that of the *conquistadores*. Slightly more capable species and civilizations may be able to eliminate the competition. But this does not mean that the competition would not eventually have developed comparable capabilities if they had been left alone.

Mayr asserts that plants do not receive "electronic" signals. By this I assume he means "electromagnetic" signals. But plants *do*. Their fundamental existence depends on receiving electromagnetic radiation from the Sun. Photosynthesis and phototropism can be found not only in the simplest plants but also in protista.

All stars emit visible light, and Sun-like stars emit most of their electromagnetic radiation in the visible part of the spectrum. Sensing light is a much more effective way of understanding the environment at some distance, certainly much more powerful than olfactory cues. It's hard to imagine a competent technical civilization that does not devote major attention to its primary means of probing the outside world. Even if they were mainly to use visible, ultraviolet, or infrared light, the physics is exactly the same for radio waves; the difference is merely a matter of wavelength.

I do not insist that the above arguments are compelling, but neither are the contrary ones. We have not witnessed the evolution of biospheres on a wide range of planets. We have not observed many cases of what is possible and what is not. Until we have had such an experience—or detected extraterrestrial intelligence—we will of course be enveloped in uncertainty.

The notion that we can, by a priori arguments, exclude the possibility of intelligent life on the possible planets of the 400 billion stars in the Milky Way has to my ears an odd ring. It reminds me of the long series of human conceits that held us to be at the center of the universe, or different not just in degree but in kind from the rest of life on Earth, or even contended that the universe was made for our benefit (Sagan, 1994). Beginning with Copernicus, every one of these conceits has been shown to be without merit.

In the case of extraterrestrial intelligence, let us admit our ignorance, put aside a priori arguments, and use the technology we are fortunate enough to have developed to try and actually find out the answer. That is, I think, what Charles Darwin—who was converted from orthodox religion to evolutionary biology by the weight of observational evidence—would have advocated.

ERNST MAYR RESPONDS

Since all we deal with are probabilities, most of them extrapolated from a sample of one, let me make a few observations in response to Carl Sagan: (1) We have no evidence that of the enormous number of Earth-like planets "each [is] stocked with enormous numbers of species." (2) There is a world of difference between plants photosynthesizing and a civilization developing the necessary theories and instrumentation for electronic communication. (3) Sagan states, "We have not witnessed the evolution of biospheres on a wide range of planets." The truth is, we have not witnessed it on a single planet outside Earth. (4) I am not talking about the possibility of extraterrestrial intelligence; I am talking about the probability of establishing it with the available means. None of Sagan's arguments has weakened my argument that it is virtually zero. This is not a conceit but a sober calculation of probabilities. The negative answer we are bound to receive will not tell us anything about the actual possibility of some extraterrestrial intelligence somewhere.

CARL SAGAN RESPONDS

I draw the tentative conclusion that other Earth-like planets have millions of species of life on them from the same data set that leads Professor Mayr to conclude that there are no extraterrestrial technical civilizations. Mayr now concedes (observation number 4 above) that there may be extraterrestrial intelligence. (Maybe even large numbers of planets inhabited by intelligent life?) But he is dubious about whether this intelligence will have developed the means for interstellar radio communication. As I have stressed, there is absolutely no compelling way to evaluate this question except by looking for interstellar radio transmissions. That is what we are doing.

APPENDIX C

THE SEARCH FOR OUR TERRESTRIAL INTELLIGENCE

In chapters 2 and 3, I examined the questions of extraterrestrial intelligence, including how the lack of any discovery might be related to the question of whether intelligence on Earth is an inevitable outcome of the evolution of life or a spurious accident inconsistent with most of evolution. The lack of any other technologically developed species on Earth, despite millions of other species' evolutionary paths, and the very short time (in geologic terms) between the advent of human intelligence and potential existential threats (including human-caused ones such as climate change and AI) indicates that there is indeed such a relation.

The question of where intelligence fits into the evolution of life, and indeed whether intelligence is an asset for a species to survive or a hindrance might be considered academic.[1] We can't conduct a lot of experimental tests, and even if we knew the answer there would be a lot of uncertainty as to what to do with it. But the question is more than academic—it is social and political—with immediate applications to all that we do. We see it manifested in the current antiscience attitude pervading social and political life.

Two antiscience movements, in particular, may indeed be crises. They are not quite existential, but both are hugely life-threatening. One is global warming (climate change), evidenced by the extreme (one might say anti-human) weather causing damage to populations around the world. The other is the spread of SARS-CoV-2 (COVID-19) and pandemic antivaccine misinformation and superstition. The two are unrelated (as far as we know) except for the common thread of having scientific solutions rejected by politically potent fringe groups that prefer conspiracy theories to the scientific method. There are examples in Earth's history of climate change destroying civilizations (for example, the Angkor Wat, Akkadian, and Mayan) and there are many examples of devastating effects of pandemics. When we expand the consideration beyond humans, species survival has often been affected by disease and by climate change. The huge difference today is the human ability to affect both climate change and the spread of disease—to exacerbate it or to compensate for it. One could view that as a test of terrestrial intelligence, and particularly as an answer to whether intelligence is an asset to species survival.

Antiscience social attitudes are a long-standing fact of human life: the rejection of our Sun-centered solar system after discoveries by Galileo and Copernicus and rejection of the evolution of species after those by Darwin are two of the most famous examples. But, compared to climate change rejection and antivaccine promotion, those earlier examples were lesser threats to life on the planet. The mortal danger to large numbers of people is far greater from ignoring climate change and ignoring vaccines than they were from clinging to belief in an Earth-centered universe or an instant formation of humankind on October 22, 4004 BC. There are other existential threats, of course—I cited them in chapter 3 when I estimated the lifetime of intelligent civilizations to enter into the Drake equation. Carl Sagan spent a large fraction of his scientific career devoted to concerns about the dangers of nuclear proliferation—its direct effect on life and its even larger indirect threat to the environment. Nuclear winter, as it came to be called, could be an even greater danger to species survival than an asteroid impact or huge volcanic eruption. But now the antiscience existential threats that result from the outright rejection of the dangers of climate change or of pandemics are in a different category. They are willful, a sort of social suicide.

Nearly existential crises often generate a wistful search for extraterrestrial intelligence and dreams of escaping home-world problems. Ronald Reagan suggested to Mikhail Gorbachev that American–Soviet nuclear cooperation would be greatly enhanced if only there were an extraterrestrial threat. Elon Musk and Jeff Bezos have off-world ambitions, wishing to create a new civilization on Mars (Musk) or in empty space (Bezos). Many news commentators cite the need for an off-world backup in case we ruin our own planet as the motivation for the entrepreneurial push for space settlements. Carl Sagan, Bruce Murray, and I cofounded The Planetary Society to support SETI. The search is a useful guide for thinking about the big questions on the nature and future of life. But, now I have come to realize we need much more a Search for Our Terrestrial Intelligence (call it SOTI). The solutions to the problems that threaten our civilization exist here on Earth, not in space, and are not from some undiscovered alien life. Ceasing the massive injection of greenhouse gases into our atmosphere in the case of climate change and vaccinating the world's population in the case of pandemics are actions to be taken by humans. Not extraterrestrials. We only need to find the terrestrial intelligence to act.

No one knows if there is any extraterrestrial intelligence. There is no evidence of it. So far, there is only one place for data about life in the universe—Earth. And on Earth, despite four-plus billion years of history, intelligence has evolved only once, and only after about 99.995 percent of that time had passed, about two hundred thousand years ago. Technological civilization only evolved in the past thousands of those years. There is a lot to suggest that intelligence (broadly defined as capable of forming civilization and developing technology) may be an accident of evolution—a lucky circumstance (for example, from an asteroid impact or some solar hyperactivity) not easily repeated. After life quickly formed on Earth, it sat around for more than two billion years (half of Earth's history) doing nothing—staying as a single cell in the more-or-less primordial ocean. A lucky, perhaps random, set of circumstances led to the evolution of mammals and then us. These include asteroid impacts and anomalous perturbations in Earth's orbit that led to different geologic eras. And, now that we are here, the very technology we develop may lead to our rather rapid extinction—that is, in thousands or tens of thousands of years, much less time than humans have been on Earth. Asteroid impact still threatens us, along with several human-made factors. It may even be that the next step in the evolution of life leads to a robotic or AI stage, without life as we know it. That is still the unforeseeable future.

In the meantime, in the foreseeable future—for our generation, our children's, and even all the foreseeable grandchildren's generations, we have a lot of reasons to worry about terrestrial intelligence, or more specifically the lack of it. Denying the real data that show the correlation between global warming and carbon-emitting industrial activity shows a lack of intelligence. Denying the data from the now hundreds of millions of vaccinated people while ignoring the threat from the unvaccinated is another example of the lack of intelligence. These are minority views among a public that generally recognizes the dangers of climate change and the unvaccinated. But collectively that minority view hinders the global terrestrial intelligence we need to find to deal with the global human future.

We have searched for extraterrestrial intelligence for sixty years with modern communications technology and for hundreds of years with telescopes monitoring the skies. Nothing has been found—not even a good hint. It is conceivable that intelligence arose but then quickly died off due to the same kind of catastrophic threats we face. Let's hope that is not a cosmic rule—and that our intelligent species will take care of itself and last a long time.

NOTES

Acknowledgments

1. Sadly, less than thirty years with Carl, who died prematurely of a blood cancer disease in 1996 (nineteen years after we formed the Society). I retired from the Society in 2010. Bruce died in 2013.

Introduction

1. As noted in the acknowledgments, I had the privilege and enjoyment of working closely with Carl Sagan for some twenty years. He had a beautiful way of expressing things—poetic and inspiring, as was well known popularly. In addition, it was, as with this comment, usually insightful.

2. Starshot is a project of the Breakthrough Initiatives with the goal of achieving interstellar flight and enabling the discovery of extraterrestrial life. The 100 Year Starship, led by Dr. Mae Jemison (see the foreword to this book), aims "to [identify and push] the radical leaps in knowledge and technology needed to achieve interstellar flight, while pioneering and transforming breakthrough applications that enhance the quality of life for all on Earth" (quoted from their mission statement, https://www.100yss.org/mission/purpose).

3. On Earth, 82 percent of our biomass (as measured in tons of carbon, since all life here is carbon based) is in plants. Fifteen percent is bacteria, and 2 percent is fungi. The remaining includes all animals, fish, birds, and us. (Yinon M. Bar-On, Rob Phillips, and Ron Milo, "The Biomass Distribution on Earth," *Proceedings of the National Academy of Sciences* 115, no. 25 [May 2018]: 6506–6511.)

Chapter 1

1. Quoted in Stephen Puryear, "Lucretius on the Existence of Other Worlds," *Stephen Puryear* (blog), July 27, 2011, https://stephenpuryear.wordpress.com/2011/07/27/lucretius-on-the-existence-of-other-worlds/.

2.   It may surprise many that the earliest Planetary Society organizers and support-ers were as much motivated by the creation of public interest representation and advancement of knowledge as they were motivated by space exploration.

3.   Quoted in Buddhistdoor Global, "Buddhistdoor View: Extending Buddhist Prac-tice to Beings Beyond Earth," *Buddhistdoor Global*, October 11, 2019, https://www .buddhistdoor.net/features/buddhistdoor-view-extending-buddhist-practice -to-beings-beyond-earth/. This is a rather relevant comment in the much more immediate context of dealing with racism and hate crimes.

4.   For simplicity, I refer to extraterrestrial intelligence as ETI. I wanted to simply call it ET in a bow to Stephen Spielberg, who introduced "E.T." to the world through his movie about an alien visitor to Earth, but decided that would be confusing when discussing simpler, not intelligent, life in the universe. Besides that, the very well-known SETI acronym is not SET—the search for extraterrestrials.

5.   For more on Aristotle and Aquinas, see Istvan Bodnar, "Aristotle's Natural Philos-ophy," *The Stanford Encyclopedia of Philosophy* (Spring 2018 Edition), ed. Edward N. Zalta (Stanford: Stanford Univ., 2018), https://plato.stanford.edu/entries/ aristotle-natphil/, and Marie I. George, "Aquinas on Intelligent Extra-Terrestrial Life," *The Thomist* 65, no. 2 (April 2001): 239–258, https://www.unav.edu/web/ ciencia-razon-y-fe/aquinas-on-intelligent-extraterrestrial-life. As for ancient astronauts, see Riley Black, "The Idiocy, Fabrications and Lies of Ancient Aliens," *Smithsonian Magazine*, May 11, 2012, https://www.smithsonianmag.com/science -nature/the-idiocy-fabrications-and-lies-of-ancient-aliens-86294030/.

6.   See, for example, Paul Tillich, *Systematic Theology* Vol. 2 (Chicago: University of Chicago Press, 1957) and Jennifer Rosato and Alan Vincelette, eds., *Extraterrestri-als in the Catholic Imagination: Explorations in Science, Science Fiction and Religion* (Newcastle upon Tyne: Cambridge Scholars Publishing, 2021).

7.   See, for example, Jeremy Kalmanofsky, "Cosmic Theology and Earthly Religion," in *Jewish Theology in Our Time: A New Generation Explores the Foundations and Future of Jewish Belief*, ed. Elliot J. Cosgrove (Woodstock, VT: Jewish Lights Pub-lishing, 2013), 23–30, and Michale Ashkenazi, "Not the Sons of Adam: Religious Responses to ETI," *Space Policy* 8, no. 4 (November 1992): 341–351.

8.   Joseph Fielding Smith, *Doctrines of Salvation* Vol. 1 (Salt Lake City: Bookcraft, 1954), 62.

9.   Qur'an, 42:29.

10.   Jafar Al-Sadiq, quoted in Nouri Sardar, "What Does Islam Say About Aliens? A Look at Quranic Verses and Hadith," *The Muslim Vibe*, November 6, 2020, https:// themuslimvibe.com/faith-islam/science/what-does-islam-say-about-aliens-a -look-at-quranic-verses-and-hadith.

11.   Bhaktivedanta, the founder of the Hare Krishna sect, stated, "In the higher planets of the material world the yogis can enjoy more comfortable and more pleasant lives for hundreds of thousands of years . . ." (excerpted from *Easy Journey to Other Planets* by A. C. Bhaktivedanta Swami Prabhupada, courtesy of the Bhaktivedanta

Book Trust International, www.Krishna.com, http://files.krishna.com/en/pdf/e
-books/Easy_Journey_to_Other_Planets.pdf).

12. From the *Numerical Discourses of the Buddha*. See, for example, Victor A. Gunase-
kara, *Basic Buddhism: A Modern Introduction to the Buddha's Teaching*, chap. 10,
https://www.buddhismtoday.com/english/buddha/Teachings/basicteaching10
.htm.

13. Spielberg was also interested in the scientific search for ETI and contributed to
The Planetary Society's private funding of SETI. His donation to The Planetary
Society for the Mega Channel Extraterrestrial Array (META) at Harvard University helped significantly—as did his personal attendance at our dedication of that
program at the Oak Ridge Observatory radio telescope in Massachusetts.

14. See appendix C, which calls for SOTI: a Search for Our Terrestrial Intelligence.

15. I accept that general consideration, since I suspect von Däniken knew his idea was
fiction, while Lowell believed he was relying on the scientific knowledge of the
time.

16. Invoking Carl Sagan's oft-repeated truism: "extraordinary claims require extraordinary evidence."

17. As one ETI enthusiast once expressed it to Carl Sagan and me when we were
not buying his argument, "Isn't science a democracy, shouldn't we treat all ideas
equally?" No—not all ideas are created equal. The particular pseudoscience idea
that got the enthusiast's attention was the "face on Mars"—seeing the representation of a human face in the natural topography on Mars under certain lighting
conditions and then concluding it was left there by a Martian ETI.

Chapter 2

1. This definition of intelligence with reference to the extraterrestrial will be used
throughout.

2. Implicitly referring to light transmission over interplanetary and interstellar distances (that is, with lasers) and skipping over flashlights and bonfires, etc.

3. It was my personal association with these projects while I was executive director
of the Society that led to my involvement in and support for SETI. Carl Sagan
and Bruce Murray were strong advocates. Steven Spielberg, on the heels of his
wonderful movie, *E.T.*, joined our board of directors and helped us fund those
projects.

4. Academician Nikolai Kardashev was a wonderfully warm, kind, and distinguished
scientist at the Soviet (later Russian) Academy of Sciences Space Research Institute, where I got to know him well over many years of my visiting the Institute and
working on cooperative projects in Mars exploration. He worked on SETI observations for many decades using large radio astronomy telescopes in the Soviet
Union and then later worked on the development of a Russian space mission,
RadioAston, also for several decades—it was finally launched in 2011. Kardashev
died in 2019.

5. I was also honored to get to know Freeman Dyson after we held a discussion, published in the *Planetary Report*, on the practicality of interstellar flight. A brilliant (genius) physicist, no one was a bigger thinker than he. He always provoked ideas and thoughts "out-of-the-box." He spent most of his life at the Institute for Advanced Study (also the home of Albert Einstein in the U.S.) at Princeton University. He participated in a one-week study I co-led at the Keck Institute for Space Studies at Caltech and at the end thanked us for giving him the opportunity to get into the real world outside his ivory tower—to which I commented, "Freeman, you are perhaps the only person in the world who could describe Caltech as the 'real world.'"

6. Mikael Flodin, "Determining Upper Limits on Galactic ETI Civilizations Transmitting Continuous Beacon Signals in the Radio Spectrum" (thesis, Royal Institute of Technology, 2019).

7. In the 1980s, a theory of "magic frequencies" was offered to predict how extraterrestrials might transmit signals. But the theory collapsed in the presence of interstellar scintillations, which distort frequencies depending on their path through the interstellar medium. Similarly, some scientists have identified locations they suggest are good candidates for ETI—but it is speculation usually based on them being like us, not a theory.

8. There are also the alien encounter reports, which go beyond mere observation. But none of these pass the credible observer test or are accompanied by any evidence that can be analyzed.

Chapter 3

1. If you don't think that is huge, look the other way and think about what we were like a hundred thousand years ago.

2. Considering here just our Galaxy, not the whole universe.

3. Personally, I think I have been quite liberal in my estimates/guesses and suspect the bottom three values in the table should be lower.

4. See Dirk Schulze-Makuch and William Bains, *The Cosmic Zoo: Complex Life on Many Worlds* (Cham: Springer, 2017).

5. Examples are adaptation to higher altitudes (less oxygen), lactase persistence (ability to digest milk), and resistance to some diseases.

6. Alternatively, perhaps not a symbiosis of humans and machines but a gradual, but total, replacement of humans by machines.

7. See Ray Kurzweil, *The Singularity is Near: When Humans Transcend Biology* (New York: Viking, 2005).

8. Yuval Noah Harari, *Homo Deus: A Brief History of Tomorrow* (New York: Harper, 2017), Kindle.

9. René Descartes, *Discourse on the Method of Rightly Conducting the Reason and Seeking Truth in the Sciences*, trans. John Veitch (Edinburgh: 1850; Project Gutenberg, 1995), part IV, https://www.gutenberg.org/files/59/59-h/59-h.htm.

10. See chapter 6.

11. The caveat "or at least in our Galaxy" is used several times in this book out of necessity, since the universe is essentially infinite (or with limits we don't know of), and with infinity anything is possible.

12. Louis Friedman, *Human Spaceflight: From Mars to the Stars* (Tucson: Univ. of Arizona Press, 2015).

Chapter 4

1. Panspermia is discussed later in this chapter.

2. Steven A. Benner, "Defining Life," *Astrobiology* 10, no. 10 (December 2010): 1021.

3. Times are all approximate.

4. If we represent the timeline on the scale of a human lifetime, for example, one hundred years, then *Homo sapiens* appeared 22.3 minutes ago.

5. Peter D. Ward and Donald Brownlee, *Rare Earth: Why Life is Uncommon in the Universe* (New York: Copernicus, 2004).

6. See Lewis Dartnell, *Origins: How the Earth Shaped Human History* (New York: Basic Books, 2019).

7. A burst of new animal species arose in the oceans 540 million years ago. What caused this Cambrian explosion is not clear—probably it was a mix of things going on in the microbe world as it evolved and the geologic conditions on Earth producing oxygen. In any case—it seems that it was at the root of the evolution of complexity.

8. A lot of the universe exists beyond what we can see. We can only see light from a distance of about 13 billion light-years (travelling here since the origin of the universe 13.8 billion years ago). But because the universe is expanding, much of it has now moved beyond that distance—as far out as 90 billion light-years. Given that, plus the unknown dark energy and dark matter, we have a lot to learn.

9. The phrase "billions and billions" is often attributed to my friend and cofounder of The Planetary Society, Carl Sagan. He said he was unaware he ever actually used that exact phrase, but no matter—even if apocryphal it represented the point he was making. And continues to do so. Carl had a beautiful way of expressing things to public audiences—a way that was evocative and explanatory (educational) simultaneously. It was also precise.

10. Although, I personally have no complaints about our Sun. It's serving us well and we are still the only life we know of.

11. In 2011, The Planetary Society attempted to test the theory of panspermia by flying samples of various microorganisms on a round trip to Phobos, the larger of Mars's moons, on a Russian mission intended to land on Phobos, collect samples, and return them to Earth. The Russians kindly agreed to carry our "Living Interplanetary Flight Experiment" (LIFE). Unfortunately, the mission failed to get out of Earth orbit and the payload remains lost in space (or perhaps the orbit decayed and burnt up in Earth's atmosphere). The experiment (LIFE) is ready to fly again on an appropriate round-trip planetary mission.

12. Or, as many believe, is it divine? Even if that were the case, it doesn't help us to form conclusions about it happening elsewhere.

13. It was a little more complicated than my oversimplification, but that is the essence.

14. How many planets? Before the invention of the telescope, there were only six planets. Then Uranus, Neptune, and Pluto were discovered and there were nine. Then, in the Space Age, we downgraded Pluto to a dwarf planet, or as the first of a new class of solar system bodies—the Kuiper belt objects (KBOs). That made it eight planets. But the planets, the KBOs, the large moons of the outer planets, and now exoplanets all qualify as new worlds.

15. A light-year is the distance light (or any electromagnetic radiation) travels in one year, about 9.5 trillion kilometers (5.9 trillion miles).

Chapter 5

1. Edward O. Wilson, *The Meaning of Human Existence* (New York: Liveright, 2014), Kindle. To be precise, Wilson's "alone" refers to the fact that, despite millions of species evolving on Earth for millions of years, only humans have become the intelligent species. He goes on to say that intelligent species in the Galaxy are rare—not likely to be present within even ten light-years of us. But in a strange bit of anthropomorphic reasoning he does argue that in the whole Galaxy there might be more species just like us with intelligence.

2. In chapter 3, the issue of existential threats limiting the lifetime of the human species was raised. It was suggested that our existence might last just tens of thousands of years. In saying "so many worlds," I think of the T-shirt slogan popular among planetary scientists, "so many worlds, so little time." If it really is only a little time, that would be depressing—but it is hard to characterize tens of thousands of years in the future for humans as too little time when we compare it to what tens of thousands of years in the past have looked like.

3. No one knows yet—we need more space missions there to sample its composition.

4. In September 2020, scientists announced a possible discovery of phosphine (a compound of hydrogen and phosphorus) in the atmosphere of Venus. Phosphine is manufactured by organisms on Earth and thus might be an indicator of life. However, the discovery is not confirmed, and whether there could be other explanations for the phosphine is not determined.

5. See, for example, Kevin Peter Hand, *Alien Oceans: The Search for Life in the Depths of Space* (Princeton: Princeton Univ. Press, 2020).

6. An informal poll of astrobiologists at a scientific meeting rated Europa and Enceladus as better candidates than Mars for finding life. However, I wonder if that speculation is influenced more by what we don't know than what we do know.

7. An average of three-plus missions per year—not unlike the European countries of the sixteenth century sending voyages to the New World.

8. But, sadly, not Russia. Two attempts at planetary missions since the fall of the Soviet Union have ended as launch failures.

9. Although the rover did a lot more science than I did.

10. I led a Mars Sample Return study in 1977 (!) at JPL, which we hoped to fly in the 1980s.

11. Obviously, I oversimplify. The navigation and protection of the spacecraft, as well as the sampling of whatever is in the plume, are all engineering challenges—but ones that can be met.

12. Another reminder of what was said in the introduction: science really isn't about getting answers—it is about forming new questions.

13. The center of mass of a star system is called the barycenter and is determined by the combined effect of the massive star and the lighter planets orbiting it. It obviously moves as the planets move. In the case of our solar system, its barycenter is currently (2022) outside the Sun, by just a little bit.

14. "The Habitable Exoplanets Catalog," Planetary Habitability Laboratory, University of Puerto Rico at Arecibo, https://phl.upr.edu/projects/habitable-exoplanets-catalog.

15. In this case, a planet is defined as habitable if it has a mass of between 0.1 and 5 Earth masses and a surface temperature (deduced from the parent star's flux) and pressure (theoretically calculated from the planet's mass and radius estimates) able to sustain water. This is a narrow definition based on what we know and is subject to future modification based on what we don't now know but will learn later.

16. I am not sure why Kepler-452 b didn't make their top list—it orbits a G-type star, with a period like Earth's (good for seasons), a radius only 60 percent larger than Earth's, and a temperature only slightly cooler than Earth's. I expect that its huge distance (1,800 light-years) and likely large gravitational force put off its selection.

Chapter 6

1. An AU is the distance from Earth to the Sun. It takes light eight minutes to reach Earth. The distance to Jupiter is about 5 AU, to Pluto about 40 AU, and to the farthest-known Kuiper belt object (the dwarf planets like Pluto) varies between about 120 and 175 AU.

2. But not Alpha Centauri itself—it is not headed in that direction.

3. Using an outer planet gravity assist can increase this number to about thirteen kilometers per second.

4. Not sure why Barnard's Star was chosen as the destination, but the study was done in the 1970s before the discovery of exoplanets.

5. Other types of space sails that don't use light may be possible. An electromagnetic sail has been proposed that would use the charged particles of space as we fly through them—but preliminary calculations show it couldn't reach interstellar speeds. Maybe in the future someone will event gamma-ray or neutrino sails, but right now we can't even describe such things.

6. This is a project of the Breakthrough Initiatives, started and funded by Yuri Milner. They have two others with interstellar objectives: Breakthrough Watch to optically observe exoplanets with large telescopes, and Breakthrough Listen to support the radio search for extraterrestrial intelligence. I am on the Starshot Advisory Committee.

7. Meredith A. MacGregor et. al., "Discovery of an Extremely Short Duration Flare from Proxima Centauri Using Millimeter Through Far-ultraviolet Observations," *The Astrophysical Journal Letters* 911, 2 (April 2021), L25.

8. Bob Forward did invent a scheme to slow down with a laser sail. It involves detaching a portion of the sail and turning it around to receive reflected light from the laser beam bouncing off the larger main sail. The smaller sail would then be slowed down for the encounter with the destination star system. It's clever, but hard to imagine how to engineer it with a laser sail and a tiny spacecraft.

9. Sonny White has now formed the Limitless Space Institute, a study group promoting and supporting research and education about advanced propulsion relevant to interstellar travel.

10. An excellent movie that I highly recommend, not as science but as thought-provoking entertainment. Prof. Thorne has written a sort-of physics textbook to accompany the movie, *The Science of Interstellar* (W. W. Norton, 2014), that I do recommend as science.

11. Actually, the space-time we live in is already four dimensions: three of space and one of time. It is better to say the wormhole introduces a new space-time dimension.

12. Robinson also wrote the trilogy *Red Mars* (Random House, 1992), *Green Mars* (Random House, 1993), and *Blue Mars* (Random House, 1996). It described humans going to, settling, and eventually terraforming and adapting on Mars to become a multiplanet species.

13. See also Kim Stanley Robinson, "What Will It Take for Humans to Colonize the Milky Way," *Scientific American*, January 13, 2016, https://www.scientificamerican.com/article/what-will-it-take-for-humans-to-colonize-the-milky-way1/.

14. Theorized by the mathematician John von Neumann, a mid-twentieth-century pioneer in the field of computer science.

15. Joseph L. Breeden, "Gravitational Assist via Near-Sun Chaotic Trajectories of Binary Objects," *Journal of the British Interplanetary Society* 66 (2013), 190–194.

16. After the German rocket pioneer Hermann Oberth.

17. More on this Oberth maneuver in the next chapter, which describes a mission to the solar gravity lens focus.

18. Abraham Loeb, "On the Possibility of an Artificial Origin for 'Oumuamua." Preprint, submitted October 20, 2021, https://doi.org/10.48550/arXiv.2110.15213. Avi Loeb, *Extraterrestrial: The First Sign of Intelligent Life Beyond Earth* (Boston: Houghton Mifflin Harcourt, 2021).

19. Steven J. Desch and Alan P. Jackson, "1I/'Oumuamua as an N_2 Ice Fragment of an Exo Pluto Surface II: Generation of N_2 Ice Fragments and the Origin of 'Oumuamua," *Journal of Geophysical Research: Planets*, 126, no. 5 (May 2021), e06807, https://doi.org/10.1029/2020JE006807.

Chapter 7

1. A very long baseline interferometer (two telescopes separated with a 2 AU baseline) would produce high resolution, but it could not produce an image.

2.   Much of this chapter, especially the characterization of the solar gravity lens's ability to image exoplanets, is due to my colleague Slava Turyshev.

3.   Or, if you want to let your mind wander, even cities—if there were such. As discussed in chapters 2 and 3, I don't expect any.

4.   Both figures here are courtesy Slava Turyshev and Viktor Toth.

5.   The multisyllable words mean a device that converts the decay heat of radioactive material to electricity—a not very efficient process, but for small amounts of power the inefficiency is acceptable.

6.   Konstantin Batygin and Michael E. Brown, "Evidence for a Distant Giant Planet in the Solar System," *The Astrophysical Journal* 151, no. 2 (January 2016), 22, https://iopscience.iop.org/article/10.3847/0004-6256/151/2/22.

7.   Spiralling in toward the Sun is analogous to a terrestrial sailboat tacking into the wind.

8.   IKAROS stands for Interplanetary Kite-craft Accelerated by Radiation of the Sun, with a nod to Icarus of Greek mythology, who flew too close to the Sun.

9.   Don't confuse sunlight pressure with the solar wind. The latter is a much weaker force from atomic particles coming from the Sun, whereas the former is the pure energy of solar photons travelling at the speed of light.

10.   From Artur Davoyan of the University of California, Los Angeles. He is studying advanced sail materials that can withstand close passages of the Sun and can be used for solar sails.

11.   SunVane was invented by Darren Garber of NXTRAC and Nathan Barnes of L'Garde.

12.   A metamaterial is manufactured with structures that interact with light and produce capabilities beyond those of natural materials—capabilities that include thermal properties, reflectivity, and emissivity.

13.   For example, the orbiter Mars Odyssey has just celebrated its twentieth year of orbital operations.

14.   At least two groups in the U.S. are actively conducting research and development of a nuclear fusion rocket. One is Princeton Satellite Systems, taking advantage of the long-time effort at Princeton University to build a nuclear fusion reactor. They won a NASA Innovative Advanced Concepts award to study the fusion rocket (their final report is at https://www.nasa.gov/sites/default/files/atoms/files/niac_2012_phaseii_slough_fusiondrivenrocketneclearpropulsiontagged.pdf.) The other group, called Helicity Space, is seeking investors for what they claim is a novel fusion drive that will lead to practical rockets for interstellar flight. See https://www.researchgate.net/publication/343702211_Helicity_Drive_A_Novel _Scalable_Fusion_Concept_for_Deep_Space_Propulsion.

Chapter 8

1.   Make no mistake, I am not against human exploration, but only want to acknowledge its limits. In an earlier book (*Human Spaceflight: From Mars to the Stars* [University of Arizona Press, 2015]), I described why I thought humans reaching and exploring Mars was important and justified, but at the same time limited—we would go no further. By the time humans explore Mars, the robotic and informa-

tion technologies will be bringing the planets to us. Similarly, I do not propose people should stop travelling to the corners of Earth and to our national parks and only rely on VR simulations. We can reach the corners of Earth, but not the corners of our universe.

2. See Liang Shi, Beichen Li, Changil Kim, Petr Kellnhofer, and Wojciech Matusik, "Towards Real-time Photorealistic 3D Holography with Deep Neural Networks," *Nature* 591 (March 10, 2021), 234–239, https://doi.org/10.1038/s41586-020-03152-0, and Daniel Ackerman, "Using Artificial Intelligence to Generate 3D Holograms in Real Time," *MIT News*, March 10, 2021, https://news.mit.edu/2021/3d-holograms-vr-0310.

3. Getting a bit more speculative, we can imagine tourists going to other worlds this way and even tour companies organizing voyages.

4. Hyperbolic velocity is the speed of the object on its hyperbolic trajectory as it exits (or enters from afar) the solar system.

5. This is exactly what made it possible to consider solar sailing for a Comet Halley rendezvous mission in the 1970s—Halley's orbit is retrograde and prior to that had been considered impossible to rendezvous with. See, for example, Louis Friedman, *Starsailing: Solar Sails and Interstellar Travel* (New York: J. Wiley, 1988).

6. ISOs are (mathematically) comets with an infinite period—that is, an open, hyperbolic orbit as opposed to a closed, elliptical orbit.

7. Enthusiasts who believed that private entertainment companies could finance a mission to the comet helped convince the U.S. government at the time not to fund a scientific mission. But much was learned from the other missions.

8. Friedman, *Human Spaceflight: From Mars to the Stars* (Tucson: University of Arizona Press, 2015).

9. When we formed The Planetary Society in 1979–80, many assumed we would be against human space flight. Sagan and Murray were both publicly known for their opposition to the space shuttle program (which was then being used to replace all other rocket launches, even lower-cost ones), in NASA planning. We were not anti-human, we were pro-exploration: choosing to reserve the human role (which is expensive and full of safety issues) for those ventures worth the risk, such as the goal of landing humans on Mars. Gene got that and wrote both an article for our magazine and a letter to his considerable fan base, extolling the formation of The Planetary Society and its goals. The shuttle, as opposed to the ship in *Star Trek*, was going where many had gone before—and not even as far. Ultimately, the shuttle was a terrific technological and engineering accomplishment, but its program helped reverse the course of space exploration.

10. *Star Trek's* theme was the exploration of other worlds. But it inspired a new paradigm for its pursuit. Instead of conquering worlds or repelling alien invaders, exploration was peaceful and designed to be unintrusive. It was scientific, and the selection of a younger woman of color as the communications officer was revolutionary for its time. The twenty-first-century generation might take for granted women scientists in important positions, multiracial crews representing planet

Earth, and respect for aliens and other unknowns—but such was not a given in the 1970s. Roddenberry's legacy is not just the what and where of exploration that he inspired, but also its how and why. Sadly, Ms. Nichols died in 2022 as this book was going to press.

11.   Lines 239–242 of "Little Gidding," from *Four Quartets* by T. S. Eliot. Copyright © 1936 by Houghton Mifflin Harcourt Publishing Company, renewed 1964 by T. S. Eliot. Copyright © 1940, 1941, 1942 by T. S. Eliot, renewed 1968, 1969, 1970 by Esme Valerie Eliot. Used by permission of HarperCollins Publishers. The excerpts in the following paragraph are from lines 244 and 245.

Chapter 9

1.   This table is reprinted with permission from Sara I. Walker, et al. "Exoplanet Biospheres: Future Directions," *Astrobiology* 18, no. 6 (June 2018), 779–824, https://doi.org/10.1089/ast.2017.1738.

2.   Edward W. Schwieterman, et. al., "Exoplanet Biosignatures: A Review of Remotely Detectable Signs of Life," *Astrobiology* 18, no. 6 (June 2018), 666, https://doi.org/10.1089/ast.2017.1729. This and the previous citation were part of a special issue of that journal devoted to biosignatures.

3.   Although interpreting any glint from an exoplanet as an ocean will be difficult and still uncertain as to its origin. Furthermore, even if it is identified as water, in a temperate zone, and with organics in the atmospheric spectra, that still does not prove life. It only suggests it.

4.   Engaging the public with VR holograms and AI products from other worlds might lead to a new travel industry niche—visits to other planets for science and experience, but also for a bit of leisure: ski Olympus (on Mars), surf Titan, or dive Europa—as we used to put on Planetary Society T-shirts. To these we can maybe add: sail Kepler-452 b.

Appendix A

1.   Paul Davies, *The Eerie Silence: Renewing Our Search for Alien Intelligence* (Boston: Houghton Mifflin Harcourt, 2010).

2.   Shouting in a dark forest may indeed be dangerous—it might wake up a dangerous beast or, to use two metaphors, "stir up a hornet's nest" or "wake up a sleeping dog." But it also might be the smart thing to do, as when confronting something that might be scared off (say, a coyote).

3.   We should, however, give it some time, because since 1974 the signal has only travelled forty-eight light-years.

4.   It is a nice idea to give advance notice to putative Venusians and Martians that we are sending spacecraft their way. Although I think the Soviets did not wait for a reply.

5.   Gravity assist to increase a spacecraft's velocity is sometimes thought of as getting something for nothing. That's not possible: energy must be conserved. When the spacecraft flies past Jupiter, its speed with respect to the Sun increases and that

with respect to Jupiter slows down—imperceptibly, since the ratio of the changes is the ratio of the spacecraft's mass to Jupiter's mass.

6. The Arecibo message travelled at the speed of light and thus surpassed Pioneer's distance in fewer than thirty minutes in 1974.

7. Jimmy Carter, Voyager Spacecraft Statement by the President. Online by Gerhard Peters and John T. Woolley, The American Presidency Project, https://www .presidency.ucsb.edu/node/243563.

8. Greg Pass, Daniel E. Goods, Pilar Fatás, Apoorv Khandelwal, Michael Skvarla, Noriaki Nakayamada, Karen Meech, Sonia Hernandez, and Evan Hilgeman, "Altamira Comet Proof-of-Concept" (Oral presentation, the International Astronautical Congress, Paris, France, September 20, 2022; https://iafastro.directory/ iac/paper/id/68166/summary/).

9. This alone would be a meaningful and fun project, even if we never sent it anywhere. It also could be a project to archive Earth's history and place it off-Earth, not necessarily on an interstellar probe but on the Moon or Mars, for example. Just in case we destroy ourselves.

Appendix C

1. As stated in chapter 2: "There is no evidence that intelligence helps species to survive. There is some evidence that it does not." Billions of species have survived on Earth without intelligence, and the jury is still out on the one with intelligence. (I am specifically including technology development in the definition of intelligence here).

INDEX

ABOUT THE AUTHOR

Dr. Louis Friedman cofounded The Planetary Society with the late Carl Sagan and Bruce Murray. As its executive director, he led programs advancing the exploration of Mars and the search for extraterrestrial intelligence, and he initiated the privately funded LightSail project. Previously, he led advanced programs at the NASA Jet Propulsion Laboratory. His most recent book was *Planetary Adventures: from Moscow to Mars,* a memoir of experiences in the Soviet Union and Russia pursuing international cooperation in space. He is a member of the Explorers Club and a fellow of the International Academy of Astronautics.